SELLING THE FARM

descants from a recollected past

DEBRA DI BLASI

C&R Press
Conscious & Responsible

ISBN 978-1-949540-13-0
LCCN 2019957445

C&R Press
Conscious & Responsible
crpress.org

For special discounted bulk purchases, please contact:
C&R Press sales@crpress.org
Contact info@crpress.org to book events, readings and author signings.

SELLING THE FARM

descants from a recollected past

Also by Debra Di Blasi

Today Is the Day That Will Matter: An Oral History
of the New America: #AlternativeFictions

The Jirí Chronicles

Prayers of an Accidental Nature

Drought & Say What You Like

What the Body Requires

Ugly Town: A Movie

Skin of the Sun

Dirty: Dirty (as editor)

This book is dedicated, with enduring love,
to my mother, who survived to move on.

Table of Contents

"I believe that the experience of childhood is irretrievable. All that remains, for any of us, is a headful of brilliant frozen moments, already dangerously distorted by the wisdoms of maturity."

—Penelope Lively, from *Oleander, Jacaranda: A Childhood Perceived*

"We leave something of ourselves behind when we leave a place; we stay there, even though we go away. And there are things in us that we can find again only by going back there."

—Pascal Mercier, from *Night Train to Lisbon*

"It is the unremarkable that will last."

—Larry Levis, "To a Wren on Calvary"

Preface

The old house burnt to ash. The cattle dead.
Acres sold to strangers.

What is the shape of a place no longer approachable except through memory? Perhaps it's the shape of galaxy clusters that resemble the shape of a brain's neural network that resembles the shape of the internet that likely resembles the shape of spacetime itself.

Selling the Farm: Descants from a Recollected Past forms a small part of a 4D literary cartography describing the 880 acres where I grew up and how my family and I were shaped by those acres. In the same way one sees, through the lens of a telescope, a star that's been dead for millennia, I see through the lens of memory those fecund acres that no longer exist—not, at least, as *they* or *I* existed there, then, and at every *there-then* recollected. Yet, nowhere in spacetime does a *placetime* cease existing. The star is evermore. The farm too. And the dead: my father, my sister, and so many furred and feathered creatures.

Entanglement, whether micro or macro, requires omnipresence, doesn't it. Requires the "God" we created to blame it all away. Autobiography that does not doff its hat to the pretense of autobiography is mostly fiction, of course. As in quantum physics, observing the experiment alters what is observed—in my case, the word paths and slippery interstices between *present* and *past*. Revisionism. The uncanny phenomenon of memory. By this reasoning, questioning the veracity and precision of language as it overlaps 're-collection' must thereby shift the paths.

So let's not call *Selling the Farm* an autobiography, but rather a biography of a place I happened to intersect, a place I happened *to*. Consider it an attempt to understand the nature of existence for human and non-human animals alike, to pull into the present a way of writing and living most relevant to what we know of the physical universe and what we might anticipate as truer representations of how we bear witness to any place and time, and to ourselves amid it all.

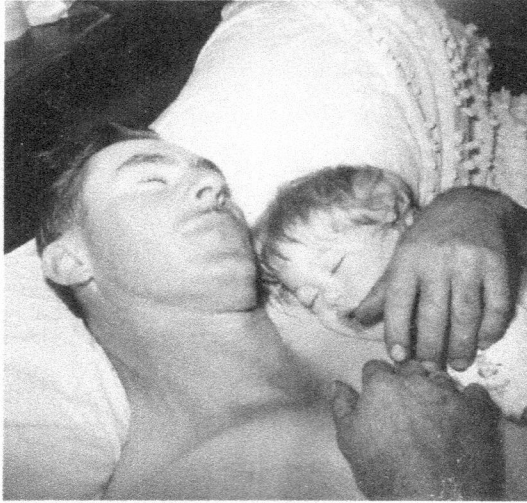

The author in the arms of her father. 1958.

The author in the arms of her sister, Diane. 1957.

In Memoriam

Donald Eugene Pickens
(1922-2017)

Diane Jean Pickens Burns
1955-2008

I will rebuild the house from the foundation up.

I will make it rise again to skies so faultless they scarcely contain
their multifarious wings.

I will walk through the front door, let it slam behind me, and
shout, I'm home!

And this time you will answer.

Overture

As I live and breathe,

> the day we moved to the farm, autumn crept up from
> the hoary lowland pastures. A capacious sky sank to gray.
> Mist rose. A chilled sun hung low and aureate on the
> overgrown lawn.

>> Well, it wasn't really a lawn, was it.
>> It was a field of weeds and grasses gone to seed.
>> Insects and moles gnawing succulent roots.
>> A garden snake slipping away. Toads. Finches.

> I was four. Third child in an oversized brood.

>> [Middle child, Tuesday's child, happy child
>> until...]

> The soft seedy foxtail grazed my jaw, and sandburs
> snagged my second-hand anklets,

as I recall.

> That ramshackle house made no sense to me who
> expected good reasons, the A + B sum.

> What kind of fool'd surface façades with crusted pebbles
> painted white cracking gray at the seams? Who'd be so
> nuts to cut 2 front doors into a 6-room house, then frame
> a warped exit to a back porch buried under rusted wheel
> rims and moldy oak planks, bent pails of coal dust and
> three-legged chairs? Who'd stick a kitchen sink and faucet
> in a place without plumbing?

And why'd the basement stairs descend to near nothing,
dank blackness fit for only salamanders and frogs, pit
under the kitchen that filled deep with a drowning rain
after big storms, and the one tiny window through which
no one could escape just a rectangle of green light choked
by outside flora gone amok?

And the sloping floors and slanted walls? The chimney full
of swallows? The knobless doors, and windows painted
shut? Nearly everything a litany of shortcuts and missteps,
things left ever undone,

as they were.

For my father would become possessed by procrastination
and piecemeal, inherit the giddy demon of half-assed
and ass-backward, makeshift and jury-rigged. So that the
house remained unplumbed, the back porch half-finished,
floors tilting like a world starting to spin off axis,

as ours would.

All flaws and failings exacerbated by time's dust and rust
and musty squalor

I so now love.

Father's recourse was to build a house up the hill a
quarter-mile distant, with running water and better views.
Then burn the old to the ground,

as he did.

Autumn

It's all there: The old house, the barns and sheds, the silo then new and shining like a dime in the distance that's not space but rather time left unhinged, a creaking door hard to open wide enough to see the whole landscape spread before me, a dream through which I continually fall. Moth at a turned season's sunrise, that's me, and daylight searing these dusty wings.

|||||

Farm objects, animate or not, pass season to season without stagger, like a deity the animals must have thought my father, moving yearlong from hay pasture to crop field, from timber to creek crossing, pausing only to inspect the light, the weather on its way. Then toss the salt blocks from the truck bed. Break the pond ice with the axe. Fill the troughs. Make life livable a sun-turn longer.

That one fine object in 880 acres of incalculable carbon organisms— *What worth beyond my recollection?* Seriously: Can we elevate the dead with language the dead can't comprehend?

A felled white oak, for example, so grand we raced upon its hundred-foot length five feet off the ground. Our pirate ship. Our hull adrift from family squalls. How long it must have lain in woodland shadows so's to gather within its slow moldering such fecund gardens of moss and fungi, skittering zoos of springtails and salamanders, red squirrels and striped skunks.

A hundred years certainly passed before that oak's trunk weakened and came down in a full-blown wind or weight of ice.

> Maybe a hundred years, too, our old house'd sat shifting on its stone foundation by the time we arrived to inhabit its wild-critter warrens. We scampered. We scavenged. We shared our hollows with rats and mice, spiders and millipedes, bees slipping through windowpanes and flies shitting on lightbulbs. And a mother and father breeding upstairs—one more child to fill one more hollow opening wide inside the mother's heart, a torn

mouth's soundless scream as dreams like dry leaves quaked and fell around her.

> [When a tree falls in the forest
> there's always someone to hear
> it crash. Some *thing* nestled in
> a darkness of its own making,
> listening.]

The house would come down around us, time after time, until at last...

But until then...

The sunset shed a pink skin on the stucco façade yearlong as long as we stayed. The six rooms shrank as we grew, suffocated us, and far away would never be far enough.

> [Until now.]

So many dead.

That moss on the oak trunk was green velvet under my hands. The scarlet cups brimmed with rainwater. Overhead, on the branch of a sapling, a nuthatch wondered at my species, invasive and wrathful all of us, ripping away old bark to expose the pill bugs beneath. The bird scolded us then flew away. My siblings stripped twigs and poked and prodded the bugs' bore holes. I looked beyond the swaying branch, at the beau-blue sky darkening toward winter slate.

I was thirteen, on the cusp of a second life. The beauty I found that day was only sadness: The oak was dead. The moss torn. The nuthatch would return, to eat the pill bugs.

|||||

"There is no object to life. To nature nothing
matters but the continuation of the species."
—W. Somerset Maugham

We wrought destruction where we could, because we could,
because we were angry about some penny-ante thing always, our
petulance a scab that could not be picked by a mother who sent
us out of the house, *out from under my hair,* she said.

Who could blame her!

The squabbling carried through a screen door slammed five times,
across the lawn and over a fence, into the woods north of the old
house where we'd search for puffballs grown roundly fat. Straining
against the carapace. Tantalizing blisters on the musky skin of
autumn earth.

> [*Our* earth, the one we owned as
> property, sure, but also possessed
> in ways we were too stupid then to
> keen, deaf as we were to premature
> nostalgia that visited and revisited
> its warning: *This, all of this, every
> sweet nuanced moment, every wild
> sense roused, every cell electric shall
> end, and you too. Do you not hear
> its Doppler shift? Do you not feel its
> long reach already stealing the pulse
> from your ticking heart?]*

We'd stomp on the puffballs to watch their spores explode from the basidiocarp. So small and manifold, those spores, they erupted as yellow or pink mist wafting out and up before settling down upon the moss and acorns, twigs and loam.

We seeded the woods that way—through rage—and thereby seeded the ensuing calm brought on by awe: that such a marvel as *puffball* existed for no other reason than to explode from raindrop or rough wind, deer hoof or girl boot.

Explode that it might multiply and thereby explode again.

Like us.

> [Like all flora and fauna
> explode and explode
> and explode.]

Like such terrifying puffballs as *We*.

| | | | |

My brother taught me how to track animals along the sandy creek banks: *This's a coyote, this's a fawn, that's a coon and that's a red squirrel...*

You could tell if the critter was heavy or light, walking or running, if it lingered at the water's edge or scampered into the brush. We'd touch a finger to where a paw had sunk in fine sand. Temperature meant time: *It was just here,* brother'd say. *Track's still warm.*

I'd look downstream to where the tracks led, hoping to see a creature careless of our scent, wobbling slow or slowly pondering its own reflection, maybe scrutinizing some pale yellow cloud of clouded sulphur butterflies at the water's edge. And I'd hope it would get away, always, from whatever hungered for it, so that each wondrous moment would rise eternal beyond recollection, a permanence more capacious than God.

My brother'd learned his craft from a book, like what he learned from all his books, the only books he kept in his room, books that in those days might have been titled, *How to Be a _Real_ Boy,* if they'd been honest about such bias. Hunting and tracking, guns and arrows, campfires and dens.

> When he was gone to woods or pastures (or later, to small-town girls and teen mischief) I'd sneak into his bedroom and study his treasures: the buck antlers and buckeyes and buckshot, the arrowheads and slabs of mica and fossils, model planes and car magazines... His world begat things redolent and inscrutable, secrets for killing and flying on land or air, and every damn lesson a prelude to escape.

He learned as well from the farm woods that were big and often gloomy, like him, a perfect habitat for what's unknown and slipping by, scarcely out of sight of what's to be dreaded: Future to come, what's not here along the beautiful creek banks but elsewhere, unhappy and without such sweet innocence as a boy seeking to be no longer such.

| | | | |

The pony named Smokey could not help herself from helping herself to summer fescue underhoof. The waxy sugary grasses tasted so fine and grew so plentiful, and since we did not corral the horses they had their reign: three hundred unfenced acres of fertile pastures at their toothy mercy.

And, mercy, that Shetland could eat!

Once, we played circus upon her ample back widened by a belly so round she always looked with foal. No reins, no harness, just a lawn of dandelions and clover and scant grass she'd munch down to the crown while I balanced atop her spine, one foot raised, ballast of arms extended. My one trick on that froward pony.

A sister sang the circus song— *Dunt-dunt duddle-uddle lunt-dunt duhn-duh...* A brother played ringmaster with a stick wand.

Flies gathered. Summer moved on.

By fall Smokey'd foundered: hooves grown long and curling back toward the coffin bones shifting. And still, she ate. She ate until the grasslands grew yellow and the weather turned chill. And when the other horses, healthy, left the high pastures for the low fallowed fields, she couldn't follow on her feverish feet.

> [A pony feels loneliness and
> thereby longing. A pony *feels*.
> You cannot tell me different.]

For weeks, now it seems in her forever haunting of me, she stood atop a hillock calling and calling to the other horses, until her whinnies became whickers, and she then went silent in an equine desolation profound as the black of her rheumy eyes.

That pony threw me the first time I tried to ride her, and every damn time after when I *giddy-upped* her flanks to turn her from ceaseless grazing. Yet I loved her more than I then loved my father who said, "There's nothing can be done," and sold her to the knackery where they'd render her corpse for dog food and soap, maybe glue from the overgrown hooves.

My last time with Smokey the sky stood gray over us. Leaves fell. Already the pond life had sunk in mud for winter, the cattails gone brittle. So much she did not understand of existence (even less than I?), like how she came to *be*, and came to be on a farm otherwise generous, and be abandoned by a herd of horses who were all she knew to be a part of, and how their absence could become another herd—stampede of loneliness that trampled her wasting body.

The knacker was coming.

I brushed the length of her spine and the ghost-white hair came away into the wind, as she was already leaving.

| | | | |

We learned to distinguish the mew from the yowl, the bark from the yelp, the grunt from squeal, whinny from whicker, bawl from bellow... Animals in pain or pleasure, happiness or fret.

Why then so difficult to rescue a person in pain?

> Eldest brother brawling, youngest brother lost in the far-off baseball field of daydreams. Eldest sister biting her folded tongue till it nearly bled, youngest sister screaming with night terrors.

> Me with my palpitations and hyperventilations, my bleeding cuticles and constricted bowels.

> Mother standing in the kitchen, next to the trashcan under the medicine cabinet, one hand full of pills, ready to leave the world, farm, husband, children, and life she thought empty as her aspirin bottle.

> Father roaring up the road in the old pickup, yelling his argument to no one but the dust blowing past and bug-splatter on the windshield, a logjam of rage so indisputable it wasn't just the sun that burned his face red.

| | | | |

Autumn begged destruction of all that spring had borne. Else next year's crop would not survive winter. Wind blew down birds' nests. Rainstorms carved gullies. Deer rutted against the buckthorn, and berries fell with semen.

We were thus: children wild in a timber wild

> [now grown to adults that seethe
> and buck against the civility
> of straight streets and squared
> corners, clothing dusted with our
> dead skin, forks and napkins,
> etiquette and rules… everything
> we censured.]

We wrote our own laws in brawls. Games played out to the end. Cheaters shamed or envied. Sore losers ostracized. We put off long as we could playing what'd be wrong elsewhere, off the farm.

We were terrible.

We pulled the wings off flying grasshoppers to make miniature geisha fans. We crushed the abdomens of lightning bugs to make jewelry that glowed in the dark. We popped ticks swollen with dog's blood, or impaled them on needles and held a match under them until they launched sputtering like tiny rockets aimed toward a farther world. A world where children grew to understand

> that every hapless thing needn't cede its right to live to a
> child's boredom;

that a winged grasshopper squatting on a single leaf of
pigweed, chewing, then stilled in that instant before it
takes flight, cancels out with a flinch of swift verdict the
parents' endless shouting inside the airless house;

that narrowing speculation to that infinitesimal moment,
that flicker, that *convulsion*, bears more resemblance to
the idea of "God" than any awful Jesus painting on your
Sunday school wall;

that a grasshopper's five eyes, the flagellum of its antennae,
saw as much heaven as we'd ever see, and more;

that an insect's disappearing whirr into a faultless sky
replicates the path of an angel fleeing in disgust...

Somehow this has become about religion.

| | | | |

The clays along the creek bed were pristine, gray, kiln perfect. And below them slabs rife with Cretaceous fossils: Mollusca sealed shut in a permanent silent sleep. Crinoid stems and cirri frozen mid-plummet.

Woodland Indians made jewelry from them: beads from sea lilies and buttons from mother-of-pearl. We collected the beads, the buttons. Then lost them. We lost them so good we did not know they were gone until

[years later, as adults around a dinner table, someone said
—*Remember Indian beads?*—

and someone else said
—*I wonder what happened to them?*—

and someone else said
—*Who knows? Who cares?*—

and we looked at that someone with a kind of sadness confirming the familial distance we all now survived against. A distance that would expand beyond this lifetime, possibly into the next, when]

we would become the dust of stars, or merely dust.

By now the creek has changed its course, its identity. The clays are hidden behind piles of tree limbs guarded by water moccasins and copperheads. The fossils remain buried in stony mud. The once unsullied waters over sand and polished pebbles run opaque always, a dirty brown not even an uncurried roan.

> [Creek shat full of pesticides and
> herbicides and fertilizers by the
> farm's new owners who cannot
> care less, really, about minnows
> tagging clear waters, who spray
> to kill weeds that won't be killed,
> only made exuberant, though
> the insects ail or die and thus the
> birds die or flee, and a land once
> symphonic goes quiet as an empty
> hall, and the ghost of my dead
> sister. And the ghosts of who we
> all were then can't find our way
> out of the purloined woods to get
> back home.]

We're lost.

Though I try so damn hard to find us.

I try: Afternoons now on another continent, at the edge of sleep, floating downstream beyond language, I believe—*vehemently*, as only a child believes—that I'm gifted with magic. I'm a conjurer, paramount of all conjurers, able to put it back just as it was then on that perfect day when:

> All of us, even our parents, sprawl on a sand bank to
> dry ourselves in sunlight shifting under trees so tall they
> disappear in summer's royal halo.

And someone's telling a story.

And someone's laughing.

And someone's found an Indian bead among the pebbles.

And the father says, *You better hang onto that. It'll be worth something someday.*

| | | | |

If you don't mind might I suggest:

> Tread tenderly upon the autumnal leaves as if they're
> pages in volumes of mounting years, and every page a
> color that cannot be salvaged from time. They break and
> tear under your heavy heel.

>> Once you were lighter, not much more than
>> spindly bones inside a watery sack. Skinny, not
>> delicate: You nearly flew as you skittered so fast
>> through the timber that you were only a lark
>> caught in the corner of a blink of a buck's eye
>> focused on sumac berries—here, to him, then
>> gone. Forgotten. Oh, maybe your fledgling scent
>> across synapses, but not savored. Maybe a trace
>> days later when the sumac berries were gone and
>> the buck lifted his head toward some sun-flicker
>> and twig-snap, then lowered his gaze to what
>> mattered.

Imagine:

> Proof you existed on that narrow cow path in autumn
> lasted no longer than the pause between gusts of wind,
> between rains or another layer of leaves. Not even a heat
> signature left for the copperheads curled among nature's
> detritus. Nothing.

> Why would mammals bother to remember you? Surely
> they sensed your softness toward them by how you stilled
> in your tracks at the sight and scent of them, and the way

you held your breath, then sighed so quietly they could not distinguish your wonder from the shifting weather.

And when all the deer turned, slowly, and trotted away, their white tails were not (as you once wrote) *flags of surrender* but rather offhand *bye-byes* like the afterthought of sleet falling behind you, or the half-considered wave of a girl from the back window of a '57 Chevy as she disappears from memory. Like the turning leaves under delicate deer hooves decayed beneath winter snows.

Therefore, I urge you:

Slow time further

[You can.]

to a fraction less than a fraction of a femtosecond, to the one infinitesimally wee moment before and the one after the heel or hoof connects with the desiccated leaf, weighs upon it, when the cellulose breaks away from the lignin and a sound happens, a music like the first vibration of a clarinet reed just as winter's long adagio begins.

Winter

Each time I dredge the farm it is first without snow. But if I force myself to think of winters there on the wide splay of land, snow looms in drifts lit by screeching sunlight, the gray days dissolved in synapses that refuse to spit up the paragon: cold wind and blustery gray skies and trees black and flesh-empty like burned bones.

| | | | |

Not in spring but when the snows fell languorous and mute and seemingly limitless upon the soft-rolling fields and horizoned swaths of leafless timber pointed like knights in formation, that's when the farm happened anew.

Flakes as big as my father's thumbnail, only clean. Fields of corn stalks softened sensual as a midlife woman's belly. Elm branches hanging snow-heavy until, now and then, a crack like a gun report echoed across the valley below the house, and we knew there'd be firewood now.

Every critter that lived beneath the kempt world kept still against the louder silence above: Foxes hunting in snow as the plangent hare's breath sang wavered, aria in an echo chamber. And hawks too, heads cocked in treetops for every scratch and itch unseen, the field mice taken so swiftly that they must have believed, had they belief, in heavenly ascent and brown-winged angels borne of sunlight and wind.

Winter erased the year's lessons. Life's classroom quieted. Blood in the veins in the ears coddled in wool.

Into that magnificent stillness I'd walk miles just to hear the absolute and great nothing/everything of it all, the lull I became before what was never the storm but rather the gentlest swelling of hope—green breath in black soil, hibernating frogs blinking awake, and my breath the loudest song, as I understood with acute astonishment I was alive.

| | | | |

When a cow died my father wrapped an iron chain around its
black bloat and dragged the corpse by tractor to the pile of old
tires that served as rubber wicks for the stiff body that seemed less
a body, less a cow now that all that made *cow* had fled the hide.

He poured gasoline into the tires and over the corpse and lit the
pyre with a single paper match. The cow no longer a cow burned
all day and night, sending up smoke signals to the cattle in the far
fields: *You too, O you too!*

In the morning all was black: the bones, the tires, the circle of earth
in the otherwise deep and merciless snow, and the winter clouds
racing east that carried no scent but the metal tang of icy wind.

I, rising earliest of five children, found my father standing inside
the burnt rim, gazing at the blackness gaping like a tunnel out
of his universe. Somber not sad. He pledged no religion but that
of placenta and maggots, hayseed and hay, and the barren spring
fields guaranteed before and above any god.

Then, or once, I saw him pick up a charred bone still warm and
stare into the nothing it would ever be.

> [Nothing for flies, nothing for
> buzzards and beetles, nothing for
> his five children who *needed*, it
> seemed, everything they did not
> really need: breakable toys and
> silly white boots and games that
> proved winners cruel.]

He tossed it away, the metacarpus, back onto the cooling hillock of skeleton and rubber. He looked at the sky and then slowly lowered his gaze to the hills a mile away where the bulls had already begun bellowing about what was theirs, and not.

|||||

You couldn't stop the winter cows from calving. Always, there'd be one born in the middle of a snowstorm. My father'd go searching the dark's whiteness for a trembling black newborn or the cow with placenta still hanging under her tail. He'd steal the calf from the bawling mother and carry it home.

So many calves in the kitchen! Legs splayed on linoleum. And piglets piled up on each other in big metal buckets set in front of the furnace vent. And kittens. And puppies. Fish and a blue parakeet.

It was a life, on that farm, of saving and killing.

And it was all right.

It was okay the way we each were borne into a system furious with incident, divining our day or night when the snows would come on blinding and no one staggering to carry us home.

| | | | |

That year winter stormed so dense and interminable that we were snowbound for five days. The gangling red snow fence disappeared under drifts gorgeous and blinding in the sunlight sigh that followed. Icicle swords threatened downward from rusted eaves to skulls. Deer lost their own scented paths and passed close by our house, leaping high to keep from sinking to antlers. Somewhere beneath, coyotes tunneled.

My mother, fretting as always over the fate of wrens, scattered bread and cornmeal over white iridescent crusts atop the dormant lawn, while my siblings and I, ordinary sparrows, tromped up the road to witness the beautiful catastrophe of that winter—*the best ever.*

> [All calamity cedes to self-
> appointed victors, and youth is
> gifted with a two-headed coin.]

We wrapped our socked feet in bread sacks to keep out the wet, and fixed our trousers inside with big rubber bands and tape. Our galoshes sank in powder that rose past our hips and sometimes higher, so that one time my baby brother disappeared entirely. And for a moment I saw his future erased of him in the blankness of that late snow. And thereafter, while he went on living, I delighted in the belief, absurd, that my pulling him from the drift had created a rift in the world—one with him, one without—so that while he grew and grew up I would glance at the parallel void beside us and appreciate, as I do yet, the reverberating depth of a simple gesture: proffered hand, and heft.

|||||

Where did the dogs go in winter?

Their water bowls covered with ice sat saturated of stillness near
the well where the iron pump stuck to ungloved skin and wind
moaned up the spout like a tuning bassoon.

We had no doghouses, perhaps because the farm lay scattered
with sheds and a vacant three-room house and a silo and big hay
barn with a loft surprisingly warm under the gently gabled roof,
musty ochre bales stacked to the rusting tin.

And I think now, maybe, unless I'm inventing memory to assuage
the guilt of some careless year or years when hormones drove me
toward vain petulance, that the dogs must have slept in dens of
scrapped lumber, inside one shed or another.

> Ah yes! The scent of animal heat so keen in my nose
> that I might now turn to find an old mutt curled in the
> corner, here in Portugal, in an apartment cleaned sterile
> of every musk it hoarded, so that, like the sudden pop of
> a void, fragrant recollection—re-collection—enters with a
> swagger to fill it all up.

> > And okay, maybe if I whistled softly through
> > my teeth as I'd learned to do then to rouse and
> > beckon the horses from the low pasture, maybe
> > that old black dog I've conjured would get up,
> > stretch himself with a dip and quiver, and come
> > to me, helicopter-tailed, expecting whatever's
> > happiness for a dog—the scratch under the chin,
> > the rubbed ears, the stroke head to tail. A whisper:

There now. There.

So many creatures I did not love as keenly as they deserved! Such short lives, all of them—and us—and only scraps of kindness tossed onto snows crusted like bones chewed cleaned of meat and marrow.

I was capable of more, wasn't I? Wasn't I the girl, death obsessed and romancing love, who'd've curled alongside in the hay a memory of how the motes climbed a ray of light from the gaping barnsides and then settled back upon the slick-haired spine of a perfectly satisfied dog.

|||||

All of the birds and foxes, the great wild mammals of those irretrievable winters, are long dead and become less than dust.

Each thousand one I never saw disappeared without fanfare or sigh.

How is it then I miss them?

|||||

The hill of the pasture in front of the house, across the shale mud road rutted or snowed under or ground to dust fine as talc, across the hog wire and barbed wire and hedge-apple fence, beyond the pile of burned tires and feed troughs empty now that the hogs had all been sold… *that* hill descended at a deadly angle between a scattering of maple trees sapping, toward and into another fenceline, rusted but too tough to give.

Of course we chose that hill for sledding.

For the steepness, sure, but maybe because it culled the meek from the bold, the stupid from the wise. You had to know how to steer the sled between the trees. Keep a straight line. Know when to jump off, and which way. You had to know how to get the longest, fastest ride without crashing.

Story of my life.

| | | | |

When did the season's first snowfall not stagger me by its glory?

As a child I sat at the window, nose to panes through which the winter wind squeezed an *et tu!* and watched the fat lazy flakes become a furious flurry into which my father disappeared: a ghost, again, headed into his netherworld of cattle and trees and acres of stubbled fields.

I believed therein amid the frozen eddying arose a pattern that held my future, some certainty that would soothe my night quaking, my day fear of . . . *something without name* . . . my shoulders always Henny-Penny tense and lifted as if to ward off an oncoming blow.

No, it's untrue.

Truly, I watched for the beauty of snow and for no other reason. Because beauty, then and just, was all enough.

| | | | |

A crow swooped down from a live branch and lit on the snow's bones.

My father hissed at it, *Ssskaw!*

The crow stood firm.

> That man squandered his poetry in stubborn silence, though I knew even then—long before he, at eighty, tried to dig the spot-on words back up —that he held names for the crow: *malice, snitch, threadle* and *sty.*

He turned his back on it that day. It was winter, one of the worst. Things to be done.

Only then did the crow fly.

|||||

In those polished days, falling snowflakes transmogrified in their
descent—from simple prisms to complex edifices so grotesque
they became beautiful. Cities in air. Ice palaces. Every winter
holiday we cut them from folded paper and strung them end-to-
end until the first and last connected.

No two alike.

[So they say.]

But who can know. Really. Who lives so long as to understand
the way a flake, like memory, mutates into citadels: ice toward the
complexity of physics, neurons against the simplicity of pain—
or rage? How the air's makeup that day changed everything:
the descent constructing angels or demons out of moments
we summon, the cherubim or putti witnessing our grand and
pathetic delusions?

Who but a god invented from fear of being God could prove
that recollecting the past distorts it, so that it alights upon the
branches, fields or rooftops. On the still and flowing waters.

| | | | |

To be first to lay down your tracks in a pristine snowplain vast and bright! Shadows inside the snowcup made of a boot.

I'd reach the stand of black-skinned elms with iced branches clicking like graceless castanets. Turn in the blue knives of shadow. See where I'd been.

The past, I saw, would dissolve in the heat of each moment, each step of the way.

Of course.

I'm almost sixty. Remembering is not being there. It's not the boot in snowfall but the shadow.

Interlude: Night

Maybe death's the farm at night without expectation of sunrise. The darkest places lost without shape or shadow: a nihility akin to the failing mind's gaps, the beyond always beyond the telescope's reach, a sleep absent dreams. Maybe it's the where right after fear and respite join hands.

‖‖‖

Last night I dreamt those we called *Indians* came back to take
what was theirs: Land without scars. Horses standing asleep
in prairies. Coyotes yapping under white oaks. Bears gnawing
marrow under rising starlight. Bison grazing.

The tribe's young chief decreed the razing of our farmhouse and
barn and fading red sheds, though the dismantling'd already
begun. I walked in floodlight. A mother suckling her infant
watched me pass as if I'd strolled from a world of ghosts she could
not possibly envision.

> I had: skinny barefoot girl in grass-stained shorts, ragged
> pullover, horn-rimmed glasses and blonde pixie cut.
> Inside dreaming I'd escaped my older self to return to me
> at ten, when I was more boy than girl—more *no one* than
> *some one*—and happy about it. Haunting Indians as they'd
> haunted me years gone in the farm's Cimmerian woods.

>> How we dreaded the Osage and Iowany ghosts
>> buried in shrouded pastures and knolls! Sensed
>> their bones sunk in deep loam below plow blades;
>> arrowheads and beads turned up and under again
>> and again, each season's corn or soybeans or alfalfa
>> weaving roots round the chert and vertebrae.

>> Fearful and shamed, we were, by fair umbrage for
>> what was stolen:

>>> [*Bounty land* of forty-acre plats doled
>>> out by Presidents Buchanan and Pierce
>>> to soldiers who killed the Indians or

each other in wars as monstrous as
those to come, these to stay.]

When big summer clouds shifted light and
shadow, an Indian ghost would appear at the
periphery amid trees with trunks so black and
towering they resembled burned pilasters of
smoky temples.

All smoke, those tales we told ourselves. Wisps of truth rising
from lying embers.

We galloped the farm's cow paths on make-believe ponies in
games no white person inhabited, wanting desperately to belong
to a tribe that would have us as we were: barefoot and wild on
bareback pintos, seeking nothing but cool water and shade, a
glimpse of wolf slipping away into a past without *Ma'unke* fences
or guns—bows and arrows fairer, perhaps, certainly more silent
than rifles, though every dying mammal's groan or squeal would
anyway scatter grackle flocks. As would the keening of the last
Indian to leave the farm, depart its woods and graveyard of
yellowing bones, just once turning to look back over a shoulder
slung low, at the rain blowing in from the future, a green
watercolor ruined.

Woodland Indians hunted rabbit and deer, fished
the creek that those days would've run full, as no
upstream dam had yet choked it off. They left
behind what fell. They left.

Dagúre rají hna je?

Words waking me from a dream wherein I did not belong. The
Chiwere language, the last native speakers dead since 1996,

haunting. Still. And yet.

Why did you come back?

I don't know.

The bones of wood buildings and metal trash and rubber toys
sink in deepening loam and loess. Time's a tumbler turned
backward. The past comes unlocked. Populated. Then the door's
barred. The moon's light dims.

I keep walking. Into a night beyond night. Chiwere voices sing
a rising and fading cacophony, descant with crickets and frogs,
coyote yips, a bobcat's scream like a mad woman's dirge. Mine.
Screaming, screaming mad, sorrowful, dreaming. Until, at last,
all native ghosts are safe to go on living inside a demolished
reassembled world.

|||||

Just around the northwest corner of the house, beyond ambit of the floodlight's glow, violent monsters lived.

[So we believed.]

And so we each and often altogether cryptically warned the sibling whose turn it was to carry the pisspot to the outhouse that night, in fair weather or foul. Our goal was to scare the scullery or scullion half to death, scare the shit out of them and, literally, out of the pot: younger ones—me too—racing past those illusory fiends, down the crumbled concrete walkway, sloshing the day's effluence from a family of seven over the red-painted rim and onto small bare feet.

More horrors crouched inside the outhouse, under the lousy corvus wings of our untempered ingenuity. Grisly things unnameable, inexorable. Devourers of naïfs and waifs, gnashers of bone, suckers of marrow. They hunkered also in the lair between outhouse and tumbledown shed aslant on cinder blocks, smack dab in the narrow where rusted hogwire kept the black cattle from our scraggy lawn.

A simple routine: Open the door. Dump the pot. *Flee!*

But if you spilled it, if you made the shit pot a shitty mess, you'd have to run back to the house and fill the pot with water pumped from the well, then run again past the pitch-black haunted corner, down the broken walk to the outhouse, and there swish the pot, dump it in one of three holes opening to smelly hell below—a pit of shit and piss, and lime to temper the pervasive stink. Mop up spillage. Wipe the rim. Toss the toilet paper. *Flee!*

Lit house to unlit outhouse & back—
Flee!—and speed's the key.

[No wonder we were fast runners,
winning track ribbons and medals:
100-yard dash, 250-yard dash, 440
and 880 relays. Secretly racing, we
were, against impending darkness.
Racing year after year from fear of
our fears.]

Listen, every damn rustle and twig-snap sent a shot of adrenalin
to our nape hairs prickling. Every sigh and snort from cattle so
black they disappeared into the aphotic woods—though the night
fairly roared with thrumming crickets, whirring katydids, whining
mosquitoes, croaking bullfrogs, hooting birds, and screaming prey
we imagined succumbing to predators whose indomitable gallop
would hunt us down, too. We delicate and delicious children
who'd likewise succumb if we could not outrun the beasts.

Yes, little daring darlings. We will succumb like
our father before us. Like our sister and cousins.
Like scraps of DNA wasted in a long mongrel
lineage.

|||||

Thereby I enter impervious night through a slant portal half a century gone.

Too clever by half.

I smell the extant devils closing in, putrid breath collective upon my neck, claws curled into me entering the wordy solute past.

|||||

From: *Here*
To: *There*

Old chum, remember that night mid-summer, after fifth grade, camping out on the farm's upland far from the old house?

> [*The Ridge*, we called it, because those unspoiled acres were the farm's highest, where a new house and trim lawn would rise eight years later and a family already teetering would finally collapse.]

Remember?

Tentless. Sleeping bags spread on wild grass and weeds. Cheap lawn chairs propped over our heads to protect us from no-name monsters we imagined sneaking up from the ravine below.

> The lawn chairs were aluminum, rickety, with cheery plastic webbing in groovy '60s colors. No fortress whatsoever, those chairs, yet a comfort against our dread.

> > Far, far away Elvis married Priscilla, and Twiggy made the cover of *Newsweek*. Girls like you—lace panties and beribboned hair—cut pictures of the Presleys from celebrity magazines. I cut my hair to resemble Twiggy's boycut. We were different.

> > [We are different.]

Dinner *al fresco* was bologna-and-cheese sandwiches on white bread slathered with mustard, tucked into a wax-paper bag. Cherry pop. Potato chips. Two homemade cookies apiece.

> Our family called bologna *moonmeat*. Later, when I slept over at your house in town, your mother asked which cold cuts I wanted, "Bologna or pimento loaf?"
>
> I pointed and said, "Moonmeat."
>
> She told me, "That's bologna!"
>
> I thought your mother (whom I loved like a second mother, a mother who'd found life livable) was criticizing my choice of meat—*bull-OWN-ee!*—so I blushed brow to breast. Panicked. I did not feel equal. Felt instead and always uncouth and penurious in your small perfect house of fine antiques and silver, thick pale carpet, damask, a bathtub and running water, private bedroom's matching furniture, attic full of the best and biggest toys.
>
> [I could go on.]

I go on:

That night the farm's endless stars embodied heaven overhead, and fireflies embodied stars, and the light in your starry eyes waxed starlit. Those lavender specs of mine, thick-lensed and cat-eyed and crooked on my crooked face, did they sparkle when I laughed? When I looked up and pointed out the Big Dipper and Orion's Belt, did my fake diamonds sparkle like stars?

All night sphinx moths spiraled. The smallest critters, no bigger than freckles, drew near the damp heat of we little girls scented with sweat from a long day wild, exploring wilderness under a perfect Missouri sun. Crickets and katydids and grasshoppers leapt. Mosquitoes, of course, drew blood. My dad came by to check on us, after he'd checked on the cattle. The old dog Rodney slept on your lap the whole damn night so that your legs went numb by morning. Owls screeched. The earth spun.

What did we talk about those endless hours that ended?

I don't remember falling asleep.

When morning broke we were covered in dew.

|||||

Brightest nights starlight swept the pond's surface, and algae resembled metallic threads loomed for a queen. Queen's ransom, those constellations, those nights and waters. They go on and on, beyond whatever small sphere of influence we hold over our lives. Beyond each past tomorrow.

|||||

Days like this

> —after a nightmare drives me back to the farm where
> evening descended so thick and dark that memory's lit
> only in haloes of lamplight or floodlight, flashlights or
> headlights—

I yearn for it all again. This time witnessed fully. Head-on. Before
a crucible of sunlight breaks over the building we named The
Little Red House.

> That three-room tumbledown filled with our
> hoard of junk my father somehow could not let
> go: boxes of rain-spoilt adventure books and cattle
> magazines, rusted iron bed, broken chair, one shoe
> and, later, the big brassy guts of a piano salvaged
> from the fire that burned the old house down to
> its cellar.

> > [The fire my mother believed my
> > father started to destroy evidence
> > of a life lived in discomfort, not
> > just the absence of plumbing but
> > of love.

> > Slanted walls and sloping floors
> > proof of a marriage gone awry.

> > And the offspring unmanageable
> > like hogs always rooting fencelines
> > to get out, though what they

needed to live was where they were.]

Now, where was I?

| | | | |

Sure, I can summon the farm into vision except what's in those
godforsaken night gaps. Recollected places so dark there's nothing
there at all. They frightened me then so much that fear's now
mapped in my neural paths, spread meandering like cow trails
through hay fields and timber.

It's why I dream we are roundly lit with floodlight on summer
nights. Beyond the edge is nothing. Or some *thing*: memories so
horrible the mind's pushed them off to the side, down the western
slope to the eroding gully shrouded in cottonwoods, and the
cottonwoods leaning more each year until they fall.

[Night falls too.]

Father knew our terror but did not understand it. He spent
long hours year-round in the oppressive dark of night fields
and timber, searching for lost cattle, checking newborns and
stillborns, the rebel boars caught in wire. If he feared, he did not
let on. Not even coyote packs scared him, nor bobcats—though
he carried a gun at times—for he was, as they said in those days,
the meanest sonofabitch in the valley of our creek, and he did not
believe in gods or devils or grisly ogres inhabiting anywhere but
a child's imagination run amok. Nope. He believed in real men
thieving and slaughtering, teenage vandals shooting and torching,
ex-cons taking shortcuts through our remote unlucky life.

Oh god yes, such darkness! Dreadful. Crumbling paradise
manufactured from interstices between lamplight and starlight.

The moon full-round merely inked the shadows already so deep
there on the isolated farm that gravity seemed tenuous: a spider's

thread tethering us to a closed universe we suspected preferred us as much dead those nights as alive. For the land held so many insects to feed, so many rumbling mammal bellies, and alar scavengers perched hopeful. Seasons tangled between famine and feast, sex and birth. Dying, last on the list, checked off in a pencil of ash, and a hale wet wind about to blow.

|||||

One night, coming home late and careless, I forgot to close the gap. The gap *between*.

> Between upper and lower acres, Father moved his black Angus from pasture to pasture, season to season. He sorted: steers from cows from bulls from heifers… numbered ear tags shuffled in a meticulous scheme that to me was all rigmarole.

The gap was not a serious gate.

> Not like the rickety wooden gate that shut strangers out of our property, out of our lives, shut us in. That gate finally hung so far off its hinges it wouldn't close proper, and my father replaced it with a fine aluminum gate that swung smoothly to and fro. No cursing required. No sweat.

The gap, instead, was makeshift and difficult. Barbed wire strung between three posts made from smallish tree branches. Rudimentary gate to be drawn taut and secured with two wire loops around a big fencepost, top and bottom. The weakest of us could not close or open it: The thick wire fought back and the barbs caught. I was not weakest but neither was I strongest. Not then. Not that day.

> Not that I even bothered. My head filled with birds at dusk, wind cooling hay fields sinking into the darkest shades of green, moonrise

and dew point. Future perfect elsewhere at a distance impossible to grasp.

> [I thought I would never look back.
> I've turned to salt. Tears and regret.
> But no. No interest in revision.
> Except as a measure of forgetting.]

That night, that one time I forgot to close the gap, Father came home late in the dark and tired from the fields. Angry. As usual. As necessary, perhaps, for I refused to appreciate the mechanisms of farm life I then considered purgatory, bleak penance for having been born. I thought it all unimportant, subordinate to my encroaching desire for platform shoes, real perfume, foreign cars and beautiful boys.

> It was all *between* that summer, wasn't it.
> Between pubescence and adolescence, nymph
> and imago, tomboy and menarche. Between
> loving the farm and trying to escape it. The
> long wide swim between simplicity and
> complexity, and flailing, flopping, trying not
> to drown.

Father let the screen door slam and stood there, just inside the living room crowded with the cursed lot of us. He glowered. Said, "Who left the goddamn gap open?"

I sank. Literally. Tried to make myself small, invisible, safe and disappeared.

The five of us siblings, like chicks pushing the weakest out of the nest, accused each other with facts and falsehoods until I was caught in the tangled truth: "I did it," I confessed, hair prickling, knees knocking on the gallows of my fretful mind.

"Get up and close it," he said.

Although I was by then nearly as tall as my father, he loomed. He filled the house with a wrath dense as that up the road, beyond the floodlight and around the bend of corrals, past the gas pump and pig lots, The Red Shed and silo, past the bull pasture and rusted out Model-T, on a pink road, mud dried to dusty clods. A quarter-mile of blind terror. Night shadows of black cattle and rooting hogs, corn rats and nighthawks, raccoons and rustling scrub insisting I had no nocturnal instincts.

I hadn't.

Didn't think I could do it. Paused at the intersection of floodlight and darkness, between known and unknown. And, my god, that walk to the gap seemed miles, hours—time and space stretching before me, receding it seemed from where I was, which was not near enough. And even arriving at the goddamned tangled gap, I wrestled overlong the weight and wire, hands shaking, huffing and moaning, near tears. And when finally the wire loop slid down and held on the fencepost, I turned and fled. Ran barefoot the length of that dark dirt road in seconds. Really. Raced into the warm fluttering light of house and home and family.

[A good story now. Cocktail chatter. Frivolous family tale. Yes?]

No.

That gap came to signify the end of one life and the beginning of another—*better*, we thought; *but not*, we learned. The old house and sheds falling down north, the new house and barn rising up south. The past contained in a self-contained family that walked up a dirt road into a fractured future and then walked away.

Except for Father, who stayed. Until he was taken away. He must have, one last night, stepped out into the grassy timbered darkness to face our demons he came to suspect were his own.

He was wrong.

They were ours.

> Dear Father:
> I'm sorry I left the gap open. If I close it
> now will we then love each other better?

|||||

The night repeats itself.

Around the corner of the old house, around the bend, up the long
rising road away, into the woods...

Revisited in memory, monsters endure and proliferate.

The night repeats itself, it does!

Yet, I'd not trade one split of terror then or now for the clue
they all insist being in times like today: sunny and blinding, my
faraway a safe empyrean before the end-all.

| | | | |

"Olbers' Paradox: Given a sky uniformly filled with stars and a light intensity falling off as the inverse square of the distance, why is the sky dark at night?"
—Jennifer Bothamley, *Dictionary of Theories*
(Gale Research International Ltd)

Nights when we looked to the north, down the deep slope, there was nothing. The nothing of dark rooms with tight-door closets and hands over eyes at the bottom of a lake. The darkest dark, wherein edges are not even the word *edge* nor its sign.

We know there's no such thing as black. Not even *way-back-when* where chilly depths of mud lakes sink bereft of all brine but summer sweat.

Imagine a boy at fifteen, middle-class, shining intellect— to be a doctor like his father, so inevitable. Then one summer day of only one like that one on the high dock above the new lake,

> [man-made by men felling acres of dark timber damming headwaters until the creek that swerved through our farm dried to less than a rivulet, a spitter, a skip, to make a lake that shouldn't be]

with all behind him history, his story, though blinding white, lay ahead in the water black-green, and sunlight a sneeze of glisten upon the whitecaps… And he—that small-town boy of *such promise*, they said—swan-diving, piercing waves and water, he

65

must have seen stars or the breath between stars when his head hit
the waterlogged stump that broke him from the neck down.

> *I can still smell the cove, the swaying reeds. That*
> *stunning blue shawl arced as the rim of my iris.*

The village beyond the farm would not recover from such
tragedies, for we possessed each other's lives wholly on at
least the knowable surface, just as waterbugs skate airless
on a skin-skim of pond until the pond's nevermore.

> *Again, there's no such thing as black, and darkness's just*
> *a slice of breadth: The fly that broke itself on last night's*
> *window is this morning dead and blackish.*

A boy's life, so inevitable, never to swim or walk again
among the razorgrass to learn pain's a moment of living in
the moment, to stretch that moment longer, near too far,
and bleed.

And so I revise:

> When he dived into the blackish water of the lake, the
> sun shattered below him as glass breaking as surely as his
> spine when he hit the sodden log beneath unseen.

> He saw stars.

In dreams, I visit places on the farm that lay hidden in darkness.
When I wake, I see stars.

> *Nothing's darkness, so is it matter—does it matter?—*
> *this long needle heated hot over flame before the fat*
> *tick of our skranking need rises missile-like on its*

boiling blood thrusters?

I've sat on a pier faced west gazing at the light on the water until it was all light—dark acquiescing to myopia and the silt puddle of daydreaming. I could, if I could, make a moment like no other go on beyond learning, yet I've given it all away. Discarded what wasn't mine.

Beloved, All:
I give you sky, a sky falling from sky, a sky filled with stars and light falling, till you ask, *Why the distant dark, Sister? Why the night's tremble?*

Am I there?

Yes, I'm here. Here, as certainly as I am not there gazing downhill at the sprawled valley black with trees shouldering night as I am not, am I. Sisters, brothers, we navigate the wanderlust that led us further and again. Genetic. Coded. Brain matter(s) that lets me stand again against the seeping night, a bleeding girl leaning on the corroding hog wire between the rusty barbs of a hedge-apple fence.

What's the place between the come and go, the went and gone? What's the tenderest?

Father built the new house upon animal trails that'd cut eons into black loam. Deer, fox, possum, coon… Hundreds by thousands of furred and skittered and feathered. The owl's scrape of wing against starlight. The frog-kick under mere.

Later, the coyotes came nosing the night's lost virginity.

When I'd bleed monthly they sniffed outside my bedroom window, and brother's hounds caged beneath mulberry trees downhill howled and strained toward, harder, until their haunches *chanked*. I think now I thought then I could smell coyote scent: musky teeth-yellow and feral.

Meanwhile, everyone's all-insistent moon shone nights backslit blackish and pulled my sloughing blood leaking.

> [Panties always stained and family
> poor enough to not waste funds
> on things hidden.]

And if I'd rise those nights, I rose to the echo of dead stars splayed across the entropic universe, lightless of lamps— so much so that even the town miles in the skinny distance went spit out to dark.

Everything's a pulse back then: cicadas, heart, treefrogs, breath and throb. Even still the way I scratch my scalp's a rhythm akin to the scritch to say I am here to better invent me there scratching.

> [Do I recall the color of my hair?
> Do I its length or texture?]

At fourteen I'm sure I must've been sorry I wasn't some other not stunned to be as I was: scrawny legs and neck, boney wrists grasping the hogwire like reins of a horse I would neither budge nor wish to budge.

> *Reckoning's when your face shall wince beneath the*
> *stripped skin, the muscles flayed, the misshapen bones*
> *yellow in the dimmed starlight of future imperfect.*

The broken boy's was the name I recalled on waking. Rousing me from a heart racing swiftly, skipping. Maybe I was drowning under green water.

So, if so, then who're the angelic dead insisting I give them my ear, here, my last remaining place in the world remaining a bit longer. Does it even matter which *who*, which *what*? I'm of a place now where definitions're rules by those *never-question-me* parallax views from one chuffing train window: Soybean fields fallow bend past. Sun's shining from the right-side east, and I think it all means spring is coming to my soul.

If soul means what I reason.

And I reckon it must.

I reckon, too, the blackness there just now above the timbered valley dissolving. Going *gone* forever. No one will again stand at the fence to see what I and we saw: night rising while stars scratched the paint off sky.

Well, subtract then the hordes of mosquitoes hatched and all at once rising at dusk, tiny violins whining at fourth octave. Let them finally descend to settle on skin and water, to a creek deeper in every night thereafter. They'll mate and breed. And we'll as well. And the simple complexity of revolving life, evolving so slowly that not even the midge on the wrist gets flicked or noticed because—

> *I recall sweet fleetings of sows unperturbed in a mud pasture cooling themselves in cooling evening mud.*

—see here, we're all looking *forward*, aren't we, toward and beyond every horizon away from where we're revolving. All of our suppositions ass-backward, literally, asses facing east as if rug-squat in prayer. Rapture's for those who respect more intensely the milkweed leaf spun white with cocoons. The aphid pearls. God, yes, immaculate. Some memory most all'll never know.

Reckoning's when your heart shall *kuh-plugh!* behind the ribs, intestines coil tight, sacrum pinch dry the spinal fluid. When your back's bent and the walker's toppled out of reach. When *when* ends.

> [Softly, dear intrepid dead!
> Do not shatter the eggshells of memory.]

> *I cannot tell you how often I dream I'm trying to get back home.*

> *I cannot tell you because I cannot remember. I suspect, though, the texture's that that's no more chaffing my skin.*

The disappeared grace on the sore Earth, on every organism and animal... Our endless sleep.

Moths and crickets this time of night seem less seamless in the lonely kingdom of recollection. Have I evolved to lack sunlight for all time, for example, during night, then day to keep the stone-closed waters behind great stone dams where young men buckle to become their short life in artificial light?

> *Folks, if you leave your children at the barbed fence a bit while longer you'll find the water sings into a blackness unforgivable, this, our all-zigzag evenings.*

What *is* significant to night? What matter matters *now*?

Life will end. We, like a broken boy, will disappear with sunlight fractured upon water, upon a farm, upon every one thing some sleeping night and—*Let me stay here a moment longer, please, I wish to fear less the night*—nocturnal beasts and bugs shall know time as their kingdom, and alone plants shall evolve to sing the myths of black boots breaking stems and sunbeams now vanquished.

For inimitable shall be their durable night, whether words or gestures, as they bend memory to keep Earth close-revolving precious, fearless, licked clean of time.

> *We cannot see the trees for the forest. It's possible, Folks, we never could.*

Summer

It was difficult, in that light, for anyone to be happy. The sky paused on the cusp of a heartbreaking cerulean, then tumbled headlong into coral. I, a romantic, felt color as if it were animal fur, soft and vital. Sunsets bruised like gut punches. The black smoke of treelines branded the soles of my feet, scalded me—the better part that believed in a god earthbound in every molecule and every space between. The ubiquity of a beauty in motion as to fully become love.

| | | | |

First, the bumblebees fussing at dandelions on a damp lawn as much a weedy pasture as the idea of *lawn*.

> [We didn't care. No one visited. Or, when the accidental guest arrived—vacuum cleaner salesman or proselytizing pastor—we knew they'd never come back:
>
> The farm was not a destination. A person either inhabited the place, inhabited the *notion* of it or, having arrived, was already leaving.]

But back to the lawn, to bees weighted with yellow pollen, to dew diamonds tucked among emerald stalks and blades, and each morning's symphony of what must surely have been trillions of creatures winged and grounded, rippling outward to and beyond the timbered horizons, alive from a point in time in space wherein my gaze met the bee's triad of ocelli that, in turn, must have seen the blue of my iris as a summer sky eclipsed by its stilled reflection.

If I could collect that bee's short life and utter significance, sweep with a tiny brush of memory its pause upon the flower the instant I stopped holding my breath and breathed a child's uncorrupted waft, then I suppose, well, I'd be a god, wouldn't I. I'd be the creator of the bee *and* the bee, and the dandelion, and the blue sting as I turned to move on, fifty years ago, already knowing then that neither the past—not even the literal and melancholy sidelong glance—nor the bee could be saved.

|||||

Then the roads.

Pink shale quarried from coalmines that stripped parts of the county naked of trees and good soil and poverty. For a while. Until the mines were spent, and the miners too, who moved on or stayed and went broke, their kids hungry and unwashed, and a town once big enough became now too small.

Half-mile shaled from house to front gate. And from the gate to the highway, a half-mile more. Years of cars and trucks and tractors crushing fissile rock down to pink fineness until it was all dust.

    And a mess.

Roads ran our bane. Ruts a foot deep. The old Chevy then Dodge got stuck on one hill or another, mud up to the fenders and tires clotted with clay. My mother sat crying in the driver's seat as we kids pushed effortful the car going forward again, spinning tires spraying mud on our clean school clothes. Yet another tardy slip waiting on the principal's desk.

    Fuck.

I didn't know that word yet but wish I would've, as the spitting *F* and coughing *K* might've assuaged the crawly something inside me: disgrace and fear and the protracted rage that follows fear and disgrace.

Or not.

Not really.

Because, thinking back on who I was then,

                                        [and who I've remained]

I recall my superb talent for finding a silver lining in every shale
clod or cloud of dust:

    Well, at least I'm not dead yet.

IIIII

Again, the roads.

When it rained the pink dust turned red as if split to blood,
stratified capillaries ruptured by a pummeling June deluge, a
bloodletting between silken ripe pastures. Puddles bleeding their
leak of radon

> [that my sister believed one source if not
> *the* source of a cancer that killed her too
> early, and killed also lots of our doomed
> generation
>
> —born between 1950 and 1963—
>
> statistically too many, in a small county,
> early dead of rare cancers stemming,
> perhaps, maybe
>
> —who will know for sure, who will tell—
>
> from radioactive strip mines and water
> and roads, and radioactive fallout that
> swept northeast in currents from the
> American Southwest where atom bombs
> blew skyhigh their strontium-shit while we
>
> —all five children—
>
> were womb-bound and then youngsters
> under a windy sky and rain littered with
> chemical decay.]

What was dust, now mud: molecules bound tight to each other, and to us. When we went tromping roads in black galoshes the clay clung to the rubber and then exponentially upon itself until our feet felt encased in concrete and a single step forward turned Sisyphean.

Still, we trundled, scraping the clay off with sticks when the task of trudging became near impossible. We ascended hills, sometimes crawling on all fours, because always beyond the peak lay a moment beyond average: checkerboard snake snoozing on the shoulder. Ten-point buck leaping its height over a fence, without grunt or huff—all elegant silence. Or waddling possum caught unawares as we crested, and so falling faint, the possum, playing possum, dead but not dead in the middle of the road, and one of us, each of us, me, I recall, pressing a finger against the stiff fake corpse to feel the hardness of pretend death.

Later, at night, unable to sleep for the stifling heat of the house and my whirligig brain, I'd turn on my side and bare my teeth and pop my eyes wide and go rigid head to toe to see if I could fake my way out of whatever predators lay in wait, crouched in the sticky dark future of my untethered imagination. Until I remembered buzzards.

||||

They're vultures, really, those carrion feeders—turkey vultures—but we called them buzzards. Everyone did. The cattle farmers and pig farmers losing livestock now and then, here and there, through a hole in a fence or broken gate, and the cow or pig dying lost in the woods or sunk in some muck in the low fields. Putrefying.

> Buzzards circle the dead. Remind us: *It's getting late, kids! It'll be dark soon!*

Back then they were things of beauty. Wide-winged birds that lived off what didn't live. Black silhouettes against bright blue. And when they swung low, their silver underwings glinted like scimitars swinging over our heads.

> Did the stink of death ascend from the hot summer earth as a rotating funnel? A twister of rot-gas? The buzzards caught in its eddy, salivating already to tenderize the tougher meat? Did their pink heads—featherless, for they stick their entire baldness into a carcass to get at the prime viscera—did they prickle with primal anticipation?

Nothing preyed on buzzards but pickups speeding down two-lane highways. The birds entranced stupid by the delectable dead, caught blind to their own demise. And the guys driving those pickups? Hotheaded farm boys speeding in for the kill, racking up points against life's inequities, hesitation and haste, bad choices pushed off to the side of their one-lane road.

> In the dusk behind my father's sunken glance, buzzards circle the dying: *It's getting late. It'll*

be dark soon. He's 94. Body tormented and wheelchair-bound. Skull caving round a failing brain. His children forgotten. His nurses pinched.

I hear he watches old videos of himself walking the fields of a farm no longer his and asks, "Who's that man there?" Then pauses. Asks, "Who's those buzzards?"

|||||

After life is life. Again. Yet again.

In the northern woods, down the long slope toward the creek, cinnabar fungus grew like bloody boxer's ears upon the decay of fallen trees. Buzzards ate the fresh or rotten carrion at the edge of cornfields harvested to stalks. Hawks, too, found a way to live inside the gone: elms finally ceding to disease, leaves dropping, bark peeling, until even the roots crumbled in soil where every worm and beetle celebrated with a flinch and spasm.

> I reckon maybe once, in long-past tribes chipping away at the irrational stone between them and *civilization*, all carrion creatures were omens, wide-winged shadows foreshadows of Death's private visit, as if the knock on the door to the cave or McMansion should surprise anyone past or future.

> The present, of course, already past as I type the 't'.

I used to stand over a hot stinking corpse thick with maggots and watch the writhing—and *listen*, for one could clearly hear the calliphora larvae's desperate hunger, the labrum-smacking slurping multiplied by hundreds.

How old was I when revulsion gave way to fascination, to respect for a creature pulled into each successive second by the same impulse as I: *Live! Live until you die!*

|||||

I once rescued a young pig from a posthole dug, for some reason, in the middle of the back pasture where no fencepost would ever be interred: The ground there lacked loam, and decades of neglect nurtured only burdock and briars.

> [Some whim of Father's, no doubt. One day thinking, *A boundary here*, so that one place is separate from another, so that differences are evident.
>
> A posthole dug as proof of intent.
>
> Though.]

The hole was nearly full of water after a long chilling rain, and she was cold, the pig, sunk up to her little pink ears, snout out to catch a breeze she could not reach but which brought the scent of feed and sows' teats and food slop we threw into the hog lots.

Why was I there that day, in that place, wandering among the weeds and rusted cans of a pasture gone to shit? What fantasy was I inventing away from where I was to take me into a future that would not include the dullness of hours on a farm in the middle of another lonely summer?

I pulled the pig out and she just stood there, motionless, shocked by hypothermia. I rubbed her pink-white hirsute back, her lean sides, until she shook her head like a wet dog and trotted off on her trotters toward the hog shed to eat.

I told my father my good deed. He thanked me—maybe the only time ever—and so I felt proud and finally of value to him for the whole of that summer, never once asking myself, as I do now: What difference did it make, saving that pig? In a year she was sold to slaughter.

IIIII

It wasn't just the land that drew us, though the land of course was magnificent: redolent, swarming, gesturing in swells of wind and silence. The pond waters too, year after blazing year, unfolded manifold worlds rising in tandem with rising heat from the sucking bottom-muds.

Seven ponds in all, finally, for a family of seven.

Though surely my father dug the last without thought of its implication. Or—and besides—*seven* implied nothing except longer work in the fields, another pasture to graze, more hay to bale, a bigger garden, and the oblique hope that your five kids were smart enough to know the high cost of a broken bone or stitched wound so as to therefore take care.

| | | | |

That fractured day as the sun pulled down the shade over the cattails at the edge of the old pond, I took my father's pistol and shot a bullfrog half-hidden among the scum.

I missed.

Or rather:

I left the frog living not dead. Bullet through its fat summer belly.

And when it leapt into the last swath of sunlight—fore and aft legs reaching athwart—the yellow throat pouch ballooned some pain or warning or reflex or all.

Then it fell, the body, the croak, the day and the idea that I would ever be my elder brother, the one I esteemed, who found killing a satisfaction against unspeakable rippling rage.

Water threaded red.

Stunned, the frog and I, into gawking at each other from a cold spectrum's termini: *predator* and *prey*. My pupils dilated black into lead blue; his narrowed to black seeds afloat in splendid gold so gold it looked real gold.

And the gold of that sunset lit upon his parietal eye, the third eye, the last eye to see a world go eternal dark.

A world where the death of a frog is nothing, no *thing*, is less than the drivel of snot I wiped on the back of my

hand, my left hand, the hand without the gun, as I sank
crying on the dried mud bank at the edge of a sinking sun
and a sinking bullfrog drawing shut its eyelids over the
magnificent gold of childhood as it died.

|||||

I cannot take back the killing of a creature or who I believed myself to be then, in ways I no longer believe, one way or another, neither wrong nor irrelevant, just as for the time being I was being myself. And then wasn't.

| | | | |

Summer rewarded.

Lawn dew lingered in the long morning shadow cast by our house. Low fields lay subdued under rising fog. Dormant creatures budged. Thereby I woke early with my inamorata— farm in the creeping heather dawn, rising soliloquies of finches and larks, mourning doves lamenting the holy supremacy of their ephemeral lives.

[I believed, then as now, *time is of the essence*. Felt life's contract came with such dire term limits:

So much to see and do before my mandate expires! Yes, hurry hurry! So much to report!]

Inside rooms finally cooled to comfort, my family snored and twitched. Outside, a fine honeyed light bent toward me as if the old stucco house was a renegade planet passing in front of a star.

[Maybe it was.]

Maybe, yes, the land revealed celestial machinations precisely as they appear to all dawn's early risers. Animal reality as real as mine.

[Realer, perhaps, if creatures lack the curse of storytelling, the burden of writing to make better or worse what flatly *is* or *was*.]

Real as the colossal bird on the fence that morning.

> [Or were two birds perched
> there, scrutinizing me as if I were
> nothing more or less than a meaty
> insect?
>
> Let's say one.]

Enormous, that one brown mottled bird. And rare: yellowed
beak long and sharp as a brass stiletto, belly a fat dusty beige,
legs squat. Foreign: neither wild turkey nor hawk, buzzard nor
owl, pheasant nor quail. Something *other* I'd never seen. As *other*
as I must've been to the bird, the way it gawked at me. Quiet.
Undaunted. Curiosity exceeding fear.

I don't recall how long I stood mesmerized by avian eyes too large
for its head, too high-set, and hollow as nothing. Time slowed.
Light bent. Dire message unfolding like a morning glory:

Reee-memmm-berrrrr!

It wasn't the bird's eyes or size or dusty color that stuck with me
all these years—memory bound but mutating from *real* with each
remembrance.

> [One bird?
>
> Two.
>
> Two birds on a rough red fence.
> American woodcocks. *Scolopax
> minor*. I'm now 98% certain.
> Though a newfound fact does not

alter the intangible past, does it,
as hence the second bird never alit
or was, was it. Relegated through
writing to a peripheral flicker of
the mind that cannot be edited
back into memory as clearly as the
one handsome woodcock herein
writ large.]

No.

What remains is how that moment of surprise and fascination
came to represent a convergence of strangers—not as alien species
but rather as what's briefly conscious passing through time at a
peculiar place we never belonged to or owned. Migrating. Yet
fixed, the bird and I, in solidarity. Bending light throughout our
tiny two-planet system lit by one morning's real star, and then
moving on.

|||||

The roof shingles were green, flecked, rough. In summer I'd climb to sit on the apex watching the sun emerge from a treeline I reckoned then

[and now can't expunge]

situated south

[though east.]

Below, the earth spun a parallax view from rim to rim: horse paddock and pond, dust and dirt clods of wheel tracks to the leaning barn, descending to lowlands, and everywhere else the sweep of long grasses, trees and sky, the viney weeds and scrub a scribbled afterthought.

But I never did.

Nope.

Never sat on the roof in morning. Not any morning in any season except the perfect still-formed summer of recollection that licenses me to become a bird at sunrise, scanning the farm and its bellows, its far caws, warbles and *to-weets*.

Because if I really can't go back to the farm

[and *really* I can *not*]

then let me return as a single bright moment perched on a green shingled roof.

Let me feel the coarseness under my skin right before something out there—spangled and tantalizing—catches a high bird's eye, and I take flight.

|||||

All over America fathers were mowing, watering, fertilizing lawns.

Ours ran to seed.

Weeds choked spare patches of bluegrass. Brome and timothy shot up ecstatic in sunshine. Dandelions seeded and reseeded themselves with each summer wind and child's careless flick, while milkweed and stinkgrass threatened that wonky square entirely to crust.

> We were, I suppose, the "cricker" stereotype: rusted cars under the floodlight and my farmer-father's red neck. Clothes wringer. Outhouse. Bucket at the well. The creek ran not proximate but near enough to suggest we were always *this close* to real lack—
>
> *Cricker!* a sinister echo rising from bottomlands.

Still, in hindsight, wasn't it best waiting weeks for a trim and a shave?

After we filled the mower's gas tank, topped up the oil, cranked the motor with a few violent yanks, after the geometry of right angles and straight lines and a good raking, the lawn looked magnificent, all tidy and just about fancy. Smelled extravagant, like what the cows and horses craved: that good-for-you lavishness lifting our eyes skyward with a long sniff and *mmmmmm*-ing sigh.

> [Yes. True enough.]

We'd go to bliss mow after mow, to a particular spot where what was heavy and depressing didn't saddle our scrawny collarbones, and richness stood redefined as green bounty in our rural-made language spilt from deep needs to speak the body's indulgence we intensely loved: chlorophyll stains on hard-calloused soles, white clover buds blent with hawkweed seed, leaf volatiles emitting their sweet S.O.S. perfume to wasps: *Come now from your papery nests under eaves, your mud clumps along barn beams! Come save us from these noisy metal blades!*

Oh, it was all so marvelous, wasn't it.

Magic:

Mow days we stayed out past sunset and moonrise, when barred owls hunted and the homely whip-poor-will choirs sang from their chancel of caliginous woods. The green grass looked black then, the raked hillock a dark velvet mountain we leapt upon, carefree and free of all the whispered judgments passed by townies.

[What did *they* know! Huh.

What *could* they know?]

In mowing, our lawn lent us a kind of prosperity

—a holy existential breadth I've not known since—

and the old house in night's light stood radiantly anew, its grass greener on both sides of the fence.

|| || |

Sliver of a day: A picnic. The whole overlarge family grazing under young mulberry trees along the rusted fence at one end of the cow pasture where the shade resided as absolution from summer's swelter.

My mother whipped open the wedding-ring quilt with its binding worn and colors faded like her marriage—except on that day, when she seemed happy for a reason unknowable that appeared rarely (and then rarer still until it vanished entirely) and she sat down upon the damp quilt spread upon a grassy patch between cow patties and paths.

She was yet young, beautiful as always, seated precisely like the nude in Manet's *Le déjeuner sur l'herbe* but clothed in denim shorts and sleeveless yellow blouse: elbow on raised knee and hand under chin, watching her children at play at the big new pond down the slope.

> [Or maybe looking past us, past all of it, toward some past or future just beyond the treeline, rewriting a story, hers, that finally turned cannibal on itself, an ouroboros that would not die but would not live, either, beyond the spinning circle of purgatory it was without end.]

Now, however, inside then and this endless tale in daylight till dusk, there's gladness drawn long as a good sigh.

Only because I make it so.

> [Keep it framed in the embellished gilt of childhood.]

So that my mother and father, amongst the bologna-and-cheese sandwiches and potato chips and grape soda, once then and ever after watch their five children race down the hill toward the mucky pond just as a sudden breeze sweeps away the paper napkins. And my mother starts to rise. And my father reaches out to lay a hand on her thigh, and squeezes, and smiles at her, and says, "Let 'em go."

| | | | |

As if a two-year-old drew a line, pretending to write longhand through timber, Blackbird Creek flowed miles through our farm. Crazy snakelike, yes, southeastward past the property line, eventually spilling into the Chariton River that spilled into the Missouri River that wound through the state until spilling into the Mississippi that flowed south to the Gulf, to the oceans.

I'd marvel. Such distance between me and a great blue whale!

And yet:

> To float face-up in early summer creek water cold and crystalline was to be threaded through the eye of the heart of the country straight to a hulking mammal skimming the sky.

> Musing, I'd skim the flowing stream with wrinkling fingertips under the celestial vault and pretend I was that quiet beast of oceans, humming to myself, wondering what's up beyond the blue between the trees, where's the moving white cumulus going?

>> [Cumulus, from Latin *to heap.*
>> And did they heap, those clouds!

>> Roiling to soundtracks of hissing leaves and bird squawks and bellowing bulls. Big-backed and silent, those clouds: *There's a buffalo, a camel, a baleen whale...*]

96

Clearly, to see eye-to-eye how clearly all things connected, micro to macro, threads enmeshed, and thus anticipate even then, as an adolescent melancholic on the cusp of puberty

[my first death of the "I" I was]

the eventual too quick destruction of everything.

Exhaling all air so that I sank submerged with the same slow sorrow of a harpooned whale hauled up in chains from salt waters and eviscerated for what we cannot stand of our own species—and what we deny—while the shared communal sky goes ever after dark over the Anthropocene's stagnant waters.

| | | | |

The friend I had was a town child, an only child. Lonely child conceived in adultery: mother not young, blood father married with children. And he—the not-kin father who chivalrously rescued them all from dubious small-town reputations—died drunk in a snowbank alongside a country road in the middle of a bitter-cold silent night when we were fourteen.

> [My friend did not know the facts
> of her conception—not then,
> and not for decades—though
> townspeople suspected all along.
> Whispered and clucked.]

And I'd say none of this—this gossipy tale I may someday rue repeating—would be important, not really, to the remembrance of *farm* and *creek*, except I'd be wrong because now, in hindsight, as her aloneness in the world seemed a ravenous scorch, a burning reach toward and toward, as if the genes of her body smelt what her mind could not decipher: the accident she was, unwanted until the godsend she became to the mother, to me in my own crowded aloneness on that farm in summer, to her own children later, and maybe even once or always to the adopting father in his untoward touching when womanhood was pressing her outward, he who might have thought lastly of her, his not-blood daughter, as he lay in the dark whiteness of his own farm snow, numb to all but a fond gratitude for the privilege of once being the old man she called *Daddy*...

I'd be wrong because I recall her, the friend I had, really

the only true friend for so many years

[And even now, though we never
speak, no friend summons in me
such melancholic affection so
that I know on my deathbed, if
I'm afforded one, she'll be one of
few I'll choose to recall from such
bygone years.]

as so damn happy there then, squatting on the warm
beige sand of our slow-flowing creek, lonely child
swaddled in big family of my siblings and I, and the
swaying arc of trees bent careful over such a small girl as
she bending over a friable castle of sand and water that—
even while a tiny infantry of minnows invaded the moat
as the castle washed away and could not, *not ever the same,*
be rebuilt—gifted her an elated long squeal of a laugh, a
chortle and squinty-eyed gleam for a perfectly perfect day
at the creek.

That friend stayed weekends on our farm in summers young and
fabulous, as she did not mind the outhouse or cold well water
or hot crowded bedrooms and slanting floors and walls, nor the
insect bites and animal corpses, the cow shit and horse shit and
pig shit, the rats in the corn silo and mice in the attic, nor the
disappearing nightscape under the gegenschein belt...

And only that friend.

Because she was trusted early on to keep the secret of how a
life others found ignoble and shame-rife can be crowned with a
kind of wealth high beyond the flat innertubes and broken dolls
of childhood. And how a covey of quail rising from cornfields

speak a particular secret name at wingtip, as write the crawdads scuttling across streambeds, and so rattle the cottonmouth snakes tangled in creek bank brush.

Only her.

Until, at 16, when we were each and all of us trying to find our peculiar name for paradise, she leaked my secret shame (my *opulence*) in a silly fit of complicity with a friend who, like so many girls those adolescent years, was also turncoat.

And so it all came to an end.

And I invited no one.

Never again.

> [And now she, my once only lonely friend whom I will always love, lives on her own farm, Midwestern remote where, I think, I *hope*, she ends each summer night satisfied, gazing at a wide ribbon of lustrous cosmic dust, with no need to remember anything but did she leave a light on somewhere behind her.]

IIIII

Summer heat radiated the sweet undulating stink of every wild thing dead and rotting and live and shitting.

Wind that seemed endless died down, and it got hot in even the fields and woods, even in waters nearly down to the bottom clays and stone. Ponds dried to cracked pavers. And the distance bled to sky, cut by such haze as if the world stood ablaze.

It's when I knew to the fevered core I was nothing more than anything else. That the brown recluse spider could rot my skin, and the copperhead poison my blood, and the stray bobcat drag me screaming into a pile of underbrush.

That they had a right, much as I, to take what was offered or put in their path.

That my knowing this did not make me immune to what would take us all someday, regardless of how or why or ways we thought ourselves too smart.

Death, like heat, the great leveler, after all. And pretending elsewise does *not* make it *not* so.

||||

It's never the new house I recall without effort, always the old—
and fondly, now that it's history and no longer an ignominious
weapon against whatever luck I found or earned.

> Funny that we waited all those years just down
> the hill to escape a place we thought abhorrent
> in summer: granddaddy longlegs clumped big as
> your head in porch corners, and flies swarming
> under your ass from the shit in the outhouse
> holes, and thin white worms swaying like cobras
> in a pungi trance, breeding in the backyard sewage
> ditch that drained from an iron pipe spilling
> straight through the wall from the kitchen sink.

> And the mice in the piano, and the pack rats in
> shoes, and raccoons on the kitchen table when the
> fish scraps hadn't been tossed after dinner.

> And humid heat so close you'd feel it wool wound
> round you at night, until finally you'd sleep from
> exhaustion.

How we edit our lives.

Nothing wrong with that.

Because for every damnable moment I wished to be elsewhere,
there exists myriad times outside of time, me so deep in the place
I was when I was right there in the midst of yet another new
thing not before witnessed: excerpt from living aimed toward
perpetual discovery, recounted now as spheres not without end

but edged with the shift I made after I'd filed in memory only
what needed filing.

And so here it is:

> A stucco house and overgrown lawn, outbuildings and
> surrounds all now gorgeous and fecund as the lilac
> bush out back that came to bend itself open as it grew
> enormous, leaving a cupola at its heart where we'd sit on
> the cool black dirt on a hot summer's day, protected from
> what we surely must have divined as a faraway place gone
> undoubtedly wholly unreachable.

|||||

Again memory's echo:

    Father lit the trash heap and walked away.

    The wind picked up.

    As he stood counting cattle on a distant hill, childhood turned to smoke. Spiraled black and gray and white across summer fields, algal waters and seasoned timber. Lifted. Then disappeared. Just like that.

        The old house afire just once but always.
        Replayed, burning, burned to wasted less.

Now within the waist-high grasses and weeds and centuries-old plantings running sweetly their own course without wounding blades, there's yet a deep square pit of what's no longer *home*. Except—

                              [as I contrive]

    —some oily scrape of burlap. Baling twine braided round a bony wrist. Maybe grain dusting watery eyes.

Spring

A farm is a microcosm that tumbles in upon itself toward perpetual reawakening. Seasons bend and deluge without simplistic demarcations in calendars. An endlessness to ends and beginnings. Rains flooding or skies withholding. Ponds iced or thick with algae, and the black loam teeming. I search for the moments in-between, when 'what is' is obscured by the obviousness of what has been or will be: the unambiguous thunderstorm, the gooseberries gone black on the bush, the swifts turning at twilight. I'm sure it's there, in the unresolved unsolved middle, where we onlookers are unmade or made better.

|||||

I have been absent.

I don't recall the ways to be exalted by, say, a plain stone in my path. Yet here it is: simple chunk of granite worn free of an aggregate boulder pushed forward by glaciers.

Millennia: The ice insisted southward. Then stopped in the valley of our farm, in the lowlands, and retreated, melting into a lake that became a river that spit down to a creek.

And a forest grew up by the creek. And the lightning burned. And the wildflowers grew from dead leaves and the fungi rose from dead trunks. And the glades and prairies took root as hyphae reached out and every root welcomed.

Centuries: The tallgrass prairies succumbed to hooves and plow. Pastures, then. Crops. And the long-legged mammals (elk and deer and wolves and big cats) and the shorter critters (rabbits and foxes, raccoons and opossums, red squirrels and field mice and chipmunks and skunks) and the innumerable winged and the finned…

They each found a paradise godless, robust, hemmed in by fencelines and no-trespass signs that kept the guns and hounds and fishhooks at bay.

Years: The old father's shoes become garbage. Gray hats burned. Bent nails. Rust. The new owners haul it all away.

Time was different then.

Now: I toss a piece of granite back into an ocean that will rise to drown us all.

| | | | |

Names of the wild plants escape me. Sailor's britches, sure, and spring beauties. Sweet William and jack-in-the-pulpit. But the others, the thousand flowers and grasses, brambles and weeds… I've forgotten so many tender words we exchanged like secret codes to Eden's gate.

There must exist between those years and these days the mind's barbed wire I ably refused to traverse. Or else a conscious half-shrugged moment when all the taxonomized life growing and fading and reawakening in woods and pastures, in culverts and along shale roads, seemed as that which no longer mattered.

Did I really wish to step across the existential fence that divided the farm from me, a *forever* expanding daily the longer I survived my own destiny?

If I'm honest:

[and I do try]

Yes, I wanted to remake myself: *Debbie* to *Debra* with a rehearsed mid-Atlantic accent and swagger and chin up. God, such pretense! I think I thought that *not* knowing well the land would pull me out of it—or it out of me—and I'd then be of bloodlines no one, not even the bluest, could fault. Too stupid, I was, to realize that by so doing I'd ceded the whole red-lettered wealth of my empire.

|||||

Spring rained heavy and interminable. The land squished. The creek rose and eddied like the great Missouri river it sought.

We piled into the pickup and plowed through mud roads to The Rock Crossing to watch its water's muddy haste and roar. It terrified me.

> [I feared drowning. (Still do.)
> Feared floods and rising waters.
> The fearsome violence of wet
> storms ripping branches from
> trees, eating rock and muck.
> Bridges gone. Rivers cresting.]

But my mother stripped down to underwear and waded into the murky rush. She grabbed hold of creekbed stones and let the torrent pull at her. And as she screamed with a girlish giggle I wondered at her buoyant happiness, how it pleased me so—even as I feared she might simply let go and float away out of our lives.

> Dearest Mother:
> We shall all be washed away. Drowning's no secret
> except to those who cannot tolerate a life without an
> *us*: We little sailors setting out toward dawn in the
> farm morning's blush, the cat's chin unscratched, the
> dog's ticks sucking, and children trying to recall your
> face beyond the blur, or the solid timbre of your voice
> gnawed by the termites of time and flint of delight:
> your angry late-forgiving eyes.

| | | | |

A rock pile grew every spring next to The Red Shed. Pink granite, black porphyry, white quartz, limestone mottled with fossilized stems and shells from a seabed long gone. Every head-sized stone hefted from farm fields, plow blades chipped or broken as my father serenading wide skies with songs of rage, echoing past cattle no longer startled, clear down to the house where I wondered not what caused the *goddamns* and *sonafabitches* but rather the man's personality growing menacing: a turbulent storm advancing with age and lost teeth.

> Hey, I was a child. Thick-headed. Born light of the world's weight, the struggles to do right by whomever judged our constant fury.

>> [By now I've sung my own ranting plenty and await the plenty more. Hindsight's a ruse, should I follow to its manufactured core.]

It grew. A foot high then three. Twenty rocks, thirty, forty... piling up over the years against The Red Shed going pink from sun bleach. I'd scour each rock for diamonds and silver, gold, rare gems like trilobites embedded in mud turned to stone. Anything worth something enough to lift us out of father's brutal fields and mother's manic madhouse.

>> Paris, I thought. Yes! Or Hollywood. Having neither seen nor been to either city except in fantasies manufactured by bad TV. Sure thing, we'd strike it rich like the Beverly Hillbillies and live happily ever after in some fine mansion,

cement pond out back, and all my fond animals
frolicking on manicured lawns. I'd wear Evening
in Paris perfume (not toilet water) and my
pajamas would be silk.

In sorting rocks I discovered histories in the scarred and broken:
encroaching glaciers that dug the valley of the creek, leaving scree
and a lake's alluvium and drift.

There were no diamonds. No silver. Not even fool's gold for
a foolhardy girl. Even the fossils were ordinary, remnants of
creatures worthless to hunters who glimpsed their quietus caught
in Missouri limestone.

But that rock pile stayed a while a palace for rats that stole corn
from the shed, until there came to be no corn, no kernels, no
shed. No father in the cornfields, and no fields.

No girl wondering at the screeching reach of time
into future tense where my hands, veined like
schist through quartz, will recall the weight of
each rock in weather—winter cold, summer hot,
spring wet, and autumn static—and will pile them
one by one upon my breast to keep me fastened to
the world a little longer.

Until I finally understand—and forget—their
long rippling silence as what was always a piercing
fare-thee-well.

|||||

Spring meant thaw. Cow patties emitted sweet grassy redolence.
Road ruts softened. Pond ice thinned to crystal transparency.

The once flown flew back, mated and warred.

Bulbs sprouted, bloomed, drew bees and lice.

Grub worms turned.

I could not run fast enough long enough. Eventually, I bled.

       I age. Am aged.

       Today's without voices adjacent light.

       Ergo, I now hear the faintest memory of May
       clouds rumbling in the distance, breeze rising,
       calves bucking.

       And in the wispy-ringed sigh between each,
       a suggestion of some horsefly or human song
       amid birth.

|||||

"Each year it happens this way, each year
Something dead comes back and lifts up its arms…"
—Charles Wright, "Looking Around"

This is not you here in my memory of me: Little girl running
barefoot on a dusty pink road toward and away.

> [You're so far gone from where you
> began that even the red threads of
> bloodline won't lead you home again.]

This is not me in my photograph of me: Big woman smiling,
teeth Photoshopped whiter, age erased with a Gaussian blur.

Not me in so many ways not me, for I am the comet passing my
own reflection in a distant star, wondering at the sun burning me
up in a tail resplendent with lit ice.

Maybe in the core of the left eye is me in you—iris so deep and dark
it's a well and the water there's cold and likely not so clean: dead
rabbit and bug wings, paramecium flourishing in a masked ball.

Not that I am your death to come but rather your plant stem
sprouting from a wasting carcass: fox fur or mole hair quivering in
a low breeze round unfurled leaves—

> [Remember.]

O to be always green and reaching upward!

To be the spring not the fall!

|||||

Aside:

Downhill tonight the Atlantic breaks high moonless waves
against a Portuguese seawall. Uphill I spend nights on
a terrace alongside the wild field where goatherds bring
bucks and does to cull spring nasturtiums and morning
glories on a slant hill no mower could navigate.

Few stars ride the celestial arc.

Toward dawn meteors scar the sky with their death
plummets, and heaven's darker than a politician's soul,
spinning broken by manufactured light—streetlamps and
neon and headlights and house lamps and the neighbor's
match as he lights another cigarette.

Well, it's all too much.

[Too little.]

Still I wait.

Sun's rising.

What did I hope to see? Some heavenly sign disproving
what I know to be true?

> [I'll never again witness the
> speckled egg of the farm's sky,
> palms of my hands lit visible
> by starlight and space dust and

the near and far bright planets'
flickering code. Nor summer
farm's scent of rising dew or fall's
falling leaves, onward into winter's
glaze and spring's thaw I've packed
away as memory fading in the
bright false light of this aging age.]

|||||

Do the trees remember us?

Days we sat in a cradle of limbs, rocking ourselves toward comfort after a lost battle or bitter draw. Self-exiled to peace at last among twigs and leaves and termites, red-headed woodpecker drumming the whole north timber hers.

The trees, did we leave them, when we left them, a salty tear, sliver of skin, scab torn away by rough bark, blood from a nail chewed so deeply it would not stop bleeding, drool pink from sucking on those bloody fingers that would not stop bleeding?

So much blood would not stop bleeding.

Even now's a hemorrhage as we siblings advance toward each other, swords drawn, hilts of accusations and secrets glinting in a furious bloody light.

The farm's sold. Trees can't be climbed. Dead wood's culled, so woodpeckers dwindle.

We're endangered.

I choose to believe somewhere inside a sweet wet ring of uncut tree trunk's an unchanged piece of who we ever were. Memento of life before the end of a family farm, end of a family. Hidden vestige a golden tree took in like an orphan, and held, and rocked with Missouri wind through succulent leaves. The tree, thereby, transformed into more than itself, as we were, once and long ago, by each other. Wherein sibling love's not nothing but DNA.

|||||

Is spring then about forgetting?

Writing these hard April rains washing away the season's pain turned to mud and crust?

Or is it just this night flying over Africa where beyond the silver wing Hadar shines brightest, insisting I become *present* and *grounded*?

Well, I no longer believe life's written in the stars.

Life's *from* the stars and *as* the stars and *of* the stars as I am stars, and I cannot feign a five-year-old's faith in signs arising from all we know we do not know.

I know: The farm-green girl into woman's gone tough. Indomitable. No more bright kites of spring spiderlings swinging like party streamers from elms to spin me fast with bliss. No more singing loud against crushing verdant spring winds. No more reaching skinny arms wide to embrace the sprouting land, the brimming waters, the extant, the all.

[Urge's quelled lest I'm suspected crazy. I'm not crazy, I'm flying. Forsaken by the shared past.]

Still, yet, halfway round the globe Hadar shines on.

I lean back begging back what's gone: every first spring.

And there were so many then.

Epilogue

It was a universe. And every bird a sun, and every louse a planet, and upon each planet nematodes gamboling like the first children of the first bipeds, single-minded, too caught up in the awe of their upright magnificence to see the yellow beak descending.

| | | | |

While I write, my father dies.

So I ask—I *insist*:

> His death shall be calm, and the light he moves toward a
> perfect dawn over the farm's eastern treeline, sun spilling
> gold and warm on alfalfa pastures and foliate cornfields,
> across rippling creek and ponds, inside corrals and sheds,
> over the big red barn—

Yes. Yes, indeed. In heart. Mind.

> —the old house echoing laughter as it peals forevermore
> in a yesterday mislaid by we-the-living, yet keenly
> formed, I say, in the dead's space outside time's spool.
> Vivacious wife and prize bull and fattest sow. And each
> child delighted, returned to gather on The Ridge, freshly
> mowed and chlorophyll-fragrant, where overhead, lavish
> fireworks explode

> > [illuminating a kingdom that came
> > and went so quickly, it now seems:
> > My graying hair. My worn teeth.]

> Where are we to say *thank you, Father*?

> > [Where are you?]

What's made us in error need hear my petition:

Beloved Father:
The cattle shall go on grazing without wooden
chutes toward slaughter. And the pigs shall loll
cooling their immortal hides in mud alongside
troughs full of sweet silage and melons.
And the horses shall gallop for the heaven
of it—electric, handsome and free. Fruitful
and multiplying, 880 acres of endless bounty
without end. Endless: timeless, spaceless.
Everywhere and nowhere every *now* and every
then here to come, on every page snow white
or spring green.

Yes, and a day without humidity. No sweat
to wipe from under your gray cotton hat,
no bottle flies biting, no swarm of gnats or
hornets, scalded neck or moldy feet. A breeze
waxing and waning with the diapason of
birds, sun-stroked legions schooling, lambent
as gold chariots, winged chariots ascending
now, arching high across the blue sky and
pullulating earth, so that you, my dead Father,
can finally witness the apogee, the sublimity
of everything devastating and divine that
broke your back and raised your hackles
and calloused your hands that sowed these
rambling descants of contrition, and love, and
nostalgia I reap for a pretty farm

let go.

Elegy

"[Wallace's Line is] a hypothetical line dividing animals
derived from Asian species on the west from those
derived from Australian species on the east."
—Jennifer Bothamley, *Dictionary of Theories*
(Gale Research International Ltd)

"The distributions of many bird species observe
[Wallace's Line], since many birds refuse to cross even
the smallest stretches of open ocean water."
—*Wikipedia*

|||||

There is no right language for grief.

They say that when a sibling dies, memories of long-forgotten moments bubble up from the tar pit of the subconscious. It is, they say, the heart exorcising the ghosts of childhood, and the mind revising the past to close yet another chapter.

They say.

And now I know they are right: My elder sister died—*how soon!*—an agonizing death from pancreatic cancer. Toward the end, her physical pain grew so intense that she was removed from home hospice, where I and my younger sister who is a nurse worked 24/7 as our sister's keepers. In the hospital she was put into a coma so as to get ahead of her exponentially increasing pain. Yet she still screamed for hours at a time— until finally, with the help of much anesthesia, her breath ceased.

My sister's cancer coincided with my husband's cancer that coincided with news reports of bee colony extinctions, declining fish and amphibian populations, shrinking ice shelves, and the gloom of so many other disasters that portend human extinction.

And so:

O heavens! Sometimes a star shines on the dirtiest thing! Why's the night dark, Little Bird? Why's the spring fall headlong into death? We don't know where we're going yet, and not a one of us has a map.

‖‖‖

There's a bird trapped in the attic.

For 18 years my sister Diane and I slept in the same bed.
We fought. We folded a line down the center with our
separate blankets. *Cross this just once and I'll…* We stepped
a line down the center of the bedroom. *Cross this ever and
you know what!*

> [If I was my sister's keeper,
> then I did not do my job.]

She was two years older and believed in God in a way I
could not endorse: Hers the god of fire and brimstone;
mine the god of wind through trees. On her deathbed big
enough for only one she leveled her gaze and asked to see
her other sister, the younger one, the sister she said was
not poisoning her.

> [If there's grace in dying,
> then living's also a lie.]

do not cross		examine
do not cross		walk
do not cross		eyes
do not cross		paths
do not cross		me
do not cross		the line

Someone must have left a door open, so...

Sing, Sister, from your feverish coma, a blue-hot awful song. Go further, the hardest further, the broken blue early call to beyond.

Sing, Sister, and fly hard. Take the serious swim in heaven's blue sea, and let it all be a tell, a bend, a last.

Sister's sundown syndrome fully descended and she plucked from the air *nothing*—that must have seemed to her superstrings vibrating between two universes—to tape to a sheet of white paper painted with *nothing* and she was happy. "How strange the watery wash," she nearly sang, "the unblued checkered trace!"

Strange, too, the blood we shared, our blue veins a linear past indigo dark as the night outside windows of overheated bedrooms. O Sister, O Sister, your sad long hopscotch hop!

Yearning cures the death of a bee. Its transparent belly and soft-tingle legs still alighting. Death wrenched me, wrenches me still. Soft's a bee, the bee your bee and mine on this stone-slab table. *O where be the vanishing line? Why be the line ever?* Night's curled as hot leaves. Whole hives die and whole species fade to behind the frontier between present and past.

Dear Beautiful Sister:
Be then the vanishing line! *Be* the line ever!

Outside a bird cries toward the vast vastness of morning into morning. The endless gawp of desert-cooked air's silent. Step between *this* and *then*, between *know* and *wince*, and squint our sun to blood veined behind our eyes. I hear you call, Sister, from the dead, a cry, a letter, a watery silence: "Yet!"

‖‖‖

…now there's a bird trapped in the attic.

O bold and salty line! Forgive our stark timidity! Unable to say *love* without cringing, or *sorry* without choking. Not even swifts could traverse the trail of saltwater of oceans and tears between us. Not even now the moon you'll not see once again. And yet, sometimes a star shines on the dirtiest thing! The ugliness of cruel and untempered childhood blanches beautiful.

Can you help me get there, Heartness? Can you help me hear me ask:

> Genes are steel, aren't they? They're iron grates scabbing the scalps of male offspring. In this way shall I not breach the line oily on such a blue, blue sea?

> [If my sister was my *thicker-than-water,* then water's air.]

Do whole worlds evolve without me and within? Meanwhile, is there always *meanwhile*?

So far, the dead do not speak except in dreams.

|||||

That bird will not stop singing!

I dream of small animals I must rescue: from tornadoes,
from floods, from fire, from starvation or disease. Last
night I dreamt you were a yellow kitten tortured by
death's appetite while I observed, tortured by your
screeling pain, yet helpless, not helping enough, not
knowing how to help, not helping myself to your rescue,
not calling, *Help help! Some fire's alighting!* And you,
scorched. And me, flushed from your bedside burning.

> [If I am my sister's keeper,
> then I kept nothing but your squall.]

Little Kitten, yellow and scabbed, I am so so sorry,
loathing my ignorance, and fear's not penance enough.
Besides, expiation's just a bootlick to the past that's not
here, no more and always. In that dream I was tired
even in sleep so I closed you in a cardboard box with
one of your own species—doomed, too, a quiet white
cat—thinking that with such companionship may there
at least be one moment of pleasure after sting before
nothing. And I handed the box across the sill to the
nameless hands in darkness that would put your yowls to
sleep out of my head.

> [If love is perfect flight,
> then you and I be severed wings.]

Yes we're winged wings of different ilk: your father of
mites and thorny locust, mine of foxtail and sap, theirs—

those sibling others—of tumbled stones and sleet. In the
scorched desert of your afterlife sparrows hop across plains
of tar and their feet are cooked off. The wind's whistle
fades to howling. Bees aren't. Nothing's righter.

> [If reprieve is night's chill and serenade,
> then it's high noon eternal.]

|||||

Listen to me, there's a bird trapped in the attic!

Ours is a logical story: *A long line of deep ocean.* Her skin sallow, her eyes shallow waters to deeper fathoms. That's the past. That's ages ago. Now love's a kind of seeing dropped upon vast waters and disappeared.

So it is always these days about the nature of *line.* Between *this* and *that, here* and *there, then* and *now, sooner* and *later.*

Child's play.

As children we promised each other to come back from the dead with an answer. What it's like. What scars rend the night or day after *after.* What's.

I've seen a blue whale's descent in every dream since your death, Sister, just a fragment, sliver of tail and the foamy slush-hush of froth folding back over itself. In the unborn sea of infinity we're less than silence: some cankered comet blazing past stars newborn or waning. Whales echo deep.

Do you swim now unfettered?

Ah, I see: Somewhere there's blue beyond my eye's iris. Somewhere, Little Sparrow, there's where you became not ever never again flanking me.

> A scrawny robin cocks her head. How deeply
> must she listen to hear worms in this parched
> desert of your dying?

| | | | |

We are that bird's lice.

Our grandmother disappeared from our helix doubled.
Left a lineage of grief and god-leaning. Every Frenchness
or Scotchness or *whatness* in her wrung out and hung dry
on the line stretched back centuries upon centuries upon
the sediments upon the beginning.

[There is no beginning.]

Isn't it really just this and these things we suffer now?
Secrets shrugged for lack of leisure, yes, that sweltered
leisure of looking back. As if what's gone now shall come
again. The father, the son, the holy you-now-ghost wing-
caught and searching.

Every bird's a tenterhook. Every wing's a sigh. The beak's
tucked out of sunlight in sleep.

and | yet

Sometimes in the night in the dark you'd wake me
with a prod from your plumpness and whisper,
Did you hear that? And the plaster in the attic
would roll down the slant of slats and you'd
whisper, *Do you hear that now?*

And I'd say, *It's just mice.*

And you'd say, *I can't sleep.*

And I'd say, *Go to sleep.*

And you'd repeat, *I can't.*

And I'd roll over knowing that you lay awake
in the big dark eyed terrified of what you did
not know.

| Your eyes are brown. | Mine are blue. |
| You told me I was adopted. | How I wished it true. |

The year we lived in the house on Patterson Street, the
year our mother went briefly mad and rose like a Gorgon
against us, I raised white mice until they overbred and
began to starve, all of them. But before, when there were
just two, I'd tuck one or the other in the front pocket of
my shirt and find you reading always and say, "I have a
question for you." And you'd cock your head listening
deeply until the mouse crawled out of my pocket and
you'd scream.

> [If I am my sister's keeper,
> then I have bent the key.]

Deathbed's a stone-slab on which you screamed, you
screamed for hours, your hands reaching for I don't
know what.

> *Here're my hands, my love, my wing.*
> *I hear you whisper, Sister, yet.*

"…yet…" she said.

The line of gaze is never vertical, is slant and

sleepy. So tell me: How did we, how could we move so far from our date of arrival? Between *maternal* and *mother* came wrought such sorrowful migration. Islands spread in the rift of brine and bend of earth yearning toward and toward and only toward. And between islands lost of light and line, so similar these species, and yet…

You are east under sunrise.　　|　　I am west over sunset.

> *Is this the little girl I carried, slight sting*
> *on the back of the knee?*

between the lavender blooms upon there the delicate bees when nothing seemed it would not last for always, even we islands remained, though isolated by deep waters of love prevented, love gone perpetually awry: that migration of species, that leading-by-the-nose final evolution of rough distant breathing.

| | | | |

Will someone please, please, please *let the bird out of the attic?*

Death's the invisible line, the you the I, the wall of glass
that can't be navigated. Birds pummel themselves to
escape what can't be flushed: lineage of women breaking
apart like beaks upon the rocks. A soft startling pirouette
of a broken-winged fall. A sister's silent gaping mouth
behest: "...*please*..."

ALCHEMY

1. Line the page of our life our cell with salt.

2. Let the salt simmer in water of streams waded
knee-deep.

3. God's a cupping overhead. Sun's light. Trees this
stand of shadow. So serious and young you in your
sea-swishing, fingertips wingtips across a blade of
refracted sun. You cock your head and squint. And
smile. No one's said a thing except now

4. I love you.

Now that the house is gone the mice have gone
and the attic's a memory of bird's nest and
all siblings aflutter and achirp. The five of us:
sparrows: plain common birds and you who's
tame, it seems, in each dream now. Tame and

craving affection, is it really you? Tiny little head!
Liceless spread of feathers! That big white dead
cat's your escort into death, and there's nothing I
can do, it seems, to delay him though I try. Tried.
I pushed closed the door, but you know as well as
I that door had no lock, no knob, just a fingerhole
through which to peer into the lamplit room pink
as your fevered cheeks when you could no longer
stand and I could no longer let you:

"Please, oh *please!*" you said. Said that, once bedridden,
death's then soon. So you pleaded, "Please, oh *please* help
me stand!"

Why these such last words caught in my craw, little bird?

> *That big white cat's your cancer*
> *inside chewing.*

> [If there's forgiveness in your eyes,
> then let me hear some starlight's singing.]

|||||

Someone left the door open, so, finally...

Listen, beloved Sister evermore, to this last night whisper:

We are *yet*. We were never *and*.

[If I am my sister's keeper,
then—

Adieu!

Why do these maps take so goddamn long?

*Shhhh, now, yes, yes, I love you,
goodbye.*

Why does grief?

Acknowledgements

To the editors and staff of C&R Press, I extend my profound appreciation for selecting *Selling the Farm* for its 2019 Nonfiction Award, a great honor that undoubtedly will help support and inspire this writer's future literary endeavors. Thank you! I am also grateful to the following book awards for their early recognition, and the publications in which works originally appeared (in some cases, in different form). Special appreciation to poet Leslie McGrath for her generous and wise suggestions regarding nuances of language and form. I am likewise indebted to journal editors who so carefully worked with me in reshaping individual descants toward clarity and lyricism. To publications that kindly offered interviews, reviews and award nominations, your support remains inspiring and humbling.

Big Other for "Elegy" (published as "Otherwise (Eulogy for Diane)" and nominating it for 2019 *Best of the Net Anthology* (Sundress Publications). February 2019.

The Collagist for "The friend I had was a town child" (published as "Turncoats") and to interviewer William Hoffacker for asking intelligent questions about the essay's origins. February 2018.

Copper Nickel for "When a cow died my father wrapped an iron chain" and "Where did the dogs go in winter," "You couldn't stop winter cows from calving," "All of the birds and foxes," "In those polished days," and "If you don't mind might I suggest" (published as "Six Descants from Winter"). October 2019.

Entropy Magazine for "Again the Roads" and "They're vultures, really" (published as "Buzzards"). January 2017.

Four Way Books for selecting an early, shorter version of *Selling the Farm* as a finalist in the 2017 Levis Prize in Poetry.

Conceptualisms: An Anthology of Prose, Poetry, Visual, Found, and Hybrid Writing as Contemporary Art (Steve Tomasula, editor; University of Alabama Press: Tallahassee, AL) for "Preface," "When a cow died my father wrapped an iron chain," "You couldn't stop the winter cows from calving," "That year winter stormed so dense," "A crow swooped down from a live branch" and "To be first to lay down your tracks." 2020.

Gravel Magazine for "One night, coming home late and careless" (published as "Gap"). January 2018.

JuxtaProse Literary Magazine for "First, the bumblebees fussing at dandelions," "As I live and breathe," and "All over America fathers were mowing" (published as "Lawn"). March 2020.

Kestrel Journal of Literature and Art for "Last night I dreamt those we called *Indians*" (published as "Dead Languages"). Fall 2017.

The Los Angeles Review for "The small white pony named Smokey" (published as "A Dead White Pony"). September 2017.

Seneca Review for selecting an early, shorter version of *Selling the Farm* as a semi-finalist in the 2017 Deborah Tall Lyric Essay Book Prize.

Triquarterly for publishing and subsequently nominating "Five Descants from a Violent Species" for *2018 Best of the Net Anthology* (Sundress Publications); they are: "Farm objects, animate or not," "We wrought destruction where we could," "Autumn begged destruction," "My brother taught me how to track animals," and "As if a two-year-old drew a line." January 2018.

Virga for "That fractured day" (published as "Bullfrog"). October 2017.

Wreckage of Reason II: XXperimental Women Writers in the 21st Century (Nava Renek, editor; Spuyten Duyvil Press: Brooklyn, NY) for "Elegy" (published as "Olbers' Paradox"). February 2014.

About the Author

Writer and artist Debra Di Blasi grew up in rural Missouri, on a livestock and crop farm. She is the author of eight books, including *Selling the Farm: Descants from a Recollected Past* which won the 2019 C&R Press Nonfiction Award; *Drought & Say What You Like* (New Directions), winner of a Thorpe Menn Literary Excellence Award; *Prayers of An Accidental Nature* (Coffee House Press); *The Jiri Chronicles* (University of Alabama Press/FC2); and *Today Is the Day That Will Matter: An Oral History of the New America: #AlternativeFictions* (Black Scat Books).

She is also a recipient of a James C. McCormick Fiction Fellowship from the Christopher Isherwood Foundation, &NOW Award for Innovative Writing, Inspiration Grant from Kansas City Metropolitan Arts Council, Cinovation/Kansas City Film Fest Screenwriting Award, and *Diagram* Innovative Fiction Award.

The film based on her novella *Drought* (directed by Lisa Moncure and cowritten by Di Blasi and Moncure) won a host of international awards and was one of only six US films invited to the 2000 *Universe Elle* section of the Cannes International Film Festival. Her writing has appeared in many journals, including *Boulevard, Chelsea, Copper Nickel, The Collagist, Entropy, The Iowa Review, Kestrel, The Los Angeles Review, New Letters, New South Fiction, Notre Dame Review, Pleiades, Triquarterly, Wayne Literary Review*, and in notable anthologies of innovative writing.

Di Blasi is a former publisher, educator and art columnist, now dividing her time between Portugal and the U.S.

C&R Press Titles

NONFICTION

Women in the Literary Landscape by Doris Weatherford, et al

Credo: An Anthology of Manifestos & Sourcebook for
Creative Writing by Rita Banerjee and Diana Norma Szokolyai

FICTION

A History Of The Cat In Nine Chapters Or Less by Anis
Shivani No Good Very Bad Asian by Leland Cheuk
Last Tower To Heaven by Jacob Paul
Surrendering Appomattox by Jacob M. Appel
Cloud Diary by Steve Mitchell
Ivy vs. Dogg by Brian Leung
While You Were Gone by Sybil Baker
Made by Mary by Laura Catherine Brown
Cloud Diary by Stevl'herbee Mitchell
Spectrum by Martin Ott

SHORT FICTION

Father of Cambodian Time-Travel Science by Braddley Bazzle
Two Californias by Robert Glick
Meditations On The Mother Tongue by An Tran
The Protester Has Been Released by Janet Sarbanes

ESSAY AND CREATIVE NONFICTION

Selling the Farm by Debra Di Blasi
In The Room Of Persistent Sorry by Kristina Marie Darling
The Internet Is For Real by Chris Campanioni
Je suis l'autre: Essays and Interrogations by Kristina Marie
Darling Immigration Essays by Sybil Baker
Death of Art by Chris Campanioni

POETRY

9 781949 540130